FEARON'S
General Science

SECOND EDITION

Lucy Jane Bledsoe

Globe Fearon Educational Publisher
Paramus, New Jersey

Paramount Publishing

Pacemaker Curriculum Advisor: Stephen C. Larsen
Stephen C. Larsen holds a B.S. and an M.S. in Speech Pathology from the University of Nebraska at Omaha, and an Ed.D. in Learning Disabilities from the University of Kansas. In the course of his career, Dr. Larsen has worked in the Teacher Corps on a Nebraska Indian Reservation, as a Fullbright senior lecturer in Portugal and Spain, and as a speech pathologist in the public schools. A full professor at the University of Texas at Austin, he has nearly twenty years' experience as a teacher trainer on the university level. He is the author of sixty journal articles, three textbooks and six widely used standardized tests including the Test of Written Learning (TOWL) and the Test of Adolescent Language (TOAL).

Subject Area Consultant: Jack Coakley
Jack Coakley holds a B.S. in Bacteriology and Immunology from the University of California, Berkeley. He is currently teaching science at El Cerrito High School, El Cerrito, California.

Editor: Stephen Feinstein
Contributing Editors: Gordy Slack, Erika Capin
Production Editor: Teresa A. Holden
Text Designer: Dianne Platner
Cover Design: Mark Ong, Side by Side Studios
Illustrations: Duane Bibby, Carolyn Reynolds
Cover Photo: © Comstock Inc.

About the Cover Photograph: Hundreds of years ago, no one knew for sure what caused lightning. Some said it was an angry message from the gods. Today we know that lightening is a static discharge of electricity between two clouds or between a cloud and the earth. To find out more about lightening and electricity, see Chapter 19.

Other photos: Lick Observatory, University of California at Santa Cruz 2, 316; FPG International Corp./Photoworld 8, 100; Steve Turner Photography 9; Ken Lucas/Biological Photo Service 26; J. R. Waaland, University of Washington/Biological Photo Service 36; AP Wide World Photos 103, 115, 125, 168, 192, 317, 326; National Aeronautics and Space Administration 194, 244; UPI/BETTMANN 261; David J. Cross/Biological Photo Service 292; Zoological Society of San Diego 48, 60; Jeroboam, Inc./Billy E. Barnes 86; Institute of Human Origins 96; San Diego Museum of Man/Dr. Eugene Boring, Chaffey College, Alta Loma, CA 106; Smithsonian Institution 152; The BETTMANN ARCHIVE, Inc. 208, 268, 280.

Other photos courtesy of: Martin L. Schneider/ Associates 6; Stanford University Visual Arts 12; Lawrence Livermore Laboratory 44; Brazilian Tourism Office 76; University of California, Department of Anthropology 70; Schwinn Corp./Wooster Magnani Advertising 118; Don Gosney Fine Flicks 128; "Ag Alert"/Publication of California Farm Bureau 138; U.S. Geological Survey/Dept. of Interior 162; Heavenly Ski Resort, Lake Tahoe, CA/NV 172; Bigge Crane & Rigging Co. 182; U.S. Windpower 220; Hewlett-Packard 230; Manuel Berriozábal, University of Texas 241; Bay Area Regional Earthquake Preparedness Project 256; Peter Radsliff 302.

Copyright © 1994 by Globe Fearon Educational Publisher, a division of Paramount Publishing, 240 Frisch Court, Paramus, New Jersey 07652. All rights reserved. No part of this book may be reproduced or transmitted in any form or by any means, electrical or mechanical, including photocopying, recording, or by any information storage and retrieval system, without permission in writing from the publisher.

ISBN 0–8224–6888–3

Printed in the United States of America

3. 10 9 8 7 6 5 4
Cover Printer/NEBC
BI

Contents

A Note to the Student	**xiii**

Unit One: What Is Science?	**1**
1: The Wonder of Science: From Atoms to Galaxies	**2**
Science Is the Study of Our Universe	4
Science Changes Constantly	4
Science Is Made of Facts and Processes	5
Take a Stand on Science	7
The Branches of Science	8
Careers in Science	8
Chapter Review	10
2: Science Skills: The Process of Discovery	**12**
The Scientific Method	14
The Metric System of Measurement	17
Metric Units of Length	18
Metric Units of Area and Volume	18
Metric Units of Weight	18
Getting a Feel for Metric	19
Science Tools and Safety	20
A Lucky Mistake	21
Chapter Review	22
Unit One Review	24

Unit Two: Life Science — 25

3: The Study of Life — 26
- Living Things Are Called Organisms — 28
- The Fields of Life Science — 28
- What Are Characteristics? — 29
- Getting and Using Food — 30
- Moving — 30
- Growing — 31
- Reproducing — 31
- Responding to the Environment — 32
- How Long Do Organisms Live? — 33
- Chapter Review — 34

4: Living Things Are Made of Cells — 36
- Matter Is Made of Elements — 38
- All Living Things Are Made of Cells — 39
- The Main Parts of a Cell — 40
- How Cells Get Energy — 42
- Differences Between Plant and Animal Cells — 43
- The Importance of DNA — 44
- Chapter Review — 46

5: The Kingdoms of Life — 48
- Classification of Organisms — 51
- Smaller Groups of Organisms — 52
- The Protist Kingdom — 54
- The Moneran Kingdom — 55
- The Fungus Kingdom — 56
- Chapter Review — 58

6: The Animal Kingdom — 60
- What Makes an Organism an Animal? — 62
- What Are Specialized Cells? — 62
- The Invertebrates — 64
- Sponges — 64
- Worms — 65
- Mollusks — 65
- Spiny-Skinned Animals — 65
- Arthropods — 66
- Insects — 66
- Crustaceans — 67
- Spiders — 67
- The Vertebrates — 68
- Fish — 68
- Amphibians — 69
- Reptiles — 70
- Birds — 71
- Mammals — 72
- Chapter Review — 74

7: The Plant Kingdom — 76
- What Are Plants? — 78
- Roots Get Water from Soil — 78
- The Stem Is the Pathway for Food and Water — 78
- Leaves Are the Food Makers — 79
- Flowers Are the Seed Makers — 81
- Seeds Are Young Plants in a Protective Coat — 82
- Fruits Are the Ovaries of Plants — 83
- Chapter Review — 84

8: Genetics: The Code of Life — 86
- What Is Heredity? — 88
- The Beginning of Genetics — 88
- Chromosomes and Genes — 90
- How an Organism Gets Its Genes — 91
- Mutations — 91

Plant and Animal Breeding	93
Environment and Traits	93
Chapter Review	94

9: Evolution — 96

Evolution Is Change Over Time	98
DNA and Evolution	100
Charles Darwin	100
Natural Selection	101
How Mutation Causes Evolutionary Changes	102
Chapter Review	104

10: The Human Body: Cells to Systems — 106

The Organization of Cells	108
The Five Senses	109
The Nervous System	112
The Support System	113
Muscles Help You Move	113
Reproduction and Growth	114
Chapter Review	116

11: The Human Energy Systems — 118

The Circulatory System	120
The Digestive System	122
The Respiratory System	124
Chapter Review	126

12: Healthy Living — 128

What Is Disease?	130
Viruses Cause Many Diseases	131
Nutrition Means Eating Well	133
Other Ways to Stay Healthy	134
Chapter Review	136

13: Living Things Depend on Each Other — **138**
- Groups of Organisms — 140
- Ecological Change — 141
- The Food Cycle — 142
- Energy Sources — 143
- The Water Cycle — 144
- The Oxygen and Carbon-Dioxide Cycle — 145
- Conservation of Natural Resources — 146
- Chapter Review — 148
- Unit Two Review — 150

Unit Three: Physical Science — **151**

14: Properties of Matter — **152**
- Physical Science — 154
- Elements — 155
- The Structure of Atoms — 155
- The Properties of Matter — 156
- States of Matter — 157
- Compounds, Mixtures, and Solutions — 158
- Chapter Review — 160

15: Energy and Change in Matter — **162**
- What Is Energy? — 164
- The Different Forms of Energy — 165
- How Heat Energy Changes Matter — 167
- Physical and Chemical Change in Matter — 167
- Chapter Review — 170

16: Force and Motion — 172
What Is Force? — 173
Gravity and Weight — 174
Friction — 176
Centripetal Force — 177
Motion and Inertia — 178
Physics and Sports — 179
Chapter Review — 180

17: Work and Machines — 182
What Is a Machine? — 184
Levers — 186
The Lever and Mechanical Advantage — 187
Pulleys — 188
Wheel and Axle — 188
Inclined Plane — 189
The First Machines — 191
Compound Machines — 191
Chapter Review — 194

18: Heat, Light, and Sound — 196
Heat and Temperature — 198
Conduction: How Heat Moves Through Solids — 199
Convection: How Heat Moves Through Liquids and Gases — 200
Radiation: How Heat Moves Through Space — 201
Many Forms of Energy Travel in Waves — 201
What Is Light? — 202
What Happens When Light Strikes an Object? — 203
Where Does Color Come From? — 205
What Is Sound? — 206
Chapter Review — 208

19: Electricity and Magnetism — 210
- Electric Charge — 212
- Static Electricity — 213
- Lightning Is a Discharge of Electrons — 214
- Electrical Currents — 215
- Electrical Circuits — 217
- What Is Magnetism? — 218
- Magnetic Fields — 219
- Chapter Review — 220

20: Energy Resources — 222
- Fossil Fuel — 223
- Nuclear Energy — 224
- Solar Energy — 227
- Hydroelectric Energy — 228
- Geothermal Energy — 229
- Winds and Tides — 229
- Chapter Review — 230

21: Computer Technology — 232
- The Parts of a Computer — 234
- How Computers Work — 236
- The Many Uses of Computers — 239
- Chapter Review — 242
- Unit Three Review — 244

Unit Four: Earth Science — 245

22: The Earth's Features — 246

- Features of the Earth — 248
- The Age and Size of the Earth — 249
- The Three Layers of Earth — 250
- Movement of the Earth — 251
- Why Are There Seasons? — 255
- Reading a Map of the World — 253
- Time Zones Around the World — 254
- Chapter Review — 256

23: The Earth's Crust — 258

- Plate Tectonics — 260
- Colliding Plates Cause Trenches and Mountains — 262
- Rubbing Plates Cause Earthquakes — 263
- Volcanoes — 264
- Rocks and Minerals — 266
- Weathering and Erosion — 267
- Chapter Review — 268

24: The Earth's Atmosphere — 270

- What Is the Atmosphere? — 272
- Layers in the Atmosphere — 272
- What Is Air Pressure? — 274
- Heating the Atmosphere — 275
- What Causes Wind? — 275
- Local Winds — 277
- Water in the Atmosphere — 278
- Clouds — 279
- Chapter Review — 280

25: Weather and Climate — 282
- What Is Weather? — 284
- Thunderstorms — 285
- Highs and Lows — 286
- Low Pressure Storms — 286
- Weather Forecasting — 289
- What Is Climate? — 290
- Three Main Types of Climate — 291
- Chapter Review — 294

26: The Earth's History — 296
- The Age of Rocks — 297
- Fossils and the History of Life — 298
- Geological Time — 299
- Precambrian Era — 299
- Paleozoic Era — 300
- Mesozoic Era — 301
- Cenozoic Era — 302
- What Will the Next Era Be Like? — 303
- Chapter Review — 304

27: The Earth's Oceans — 306
- What Is an Ocean? — 308
- Surface Ocean Currents — 309
- Undersea Ocean Currents — 310
- Ocean Waves — 312
- Seismic Sea Waves — 313
- Tides — 313
- The Floor of the Sea — 314
- Ocean Resources — 316
- Chapter Review — 318

28: Astronomy and Space Exploration   **320**
    The Universe   321
    The Solar System   322
    Planets in the Solar System   323
    The Inner Planets   325
    The Outer Planets   327
    People Explore Space   328
    Asteroids, Meteors, and Comets   329
    Chapter Review   332
    Unit Four Review   334

Glossary   337
Appendix A: Careers in Science   349
Appendix B: Hiring Institutions   350
Appendix C: Metric Conversion Chart   351
Appendix D: Table of Chemical Elements   352
Appendix E: The Five Food Groups   356
Appendix F: Minerals   357
Appendix G: Vitamins   358
Index   361

A Note to the Student

Science is shrinking the world. Just think about it.

One hundred and fifty years ago there were no cars or trains. It took months for people to get across the United States. Then train tracks were laid across the continent. Suddenly, the country became smaller. A traveler could get from New York to San Francisco in a matter of days.

Next the automobile brought people closer together. Then television began showing us how people all over the world lived. Airplanes began being used in the early part of the 20th century. And today, computers can send messages anywhere in split seconds. Computers give answers to complicated problems before anyone can bat an eye.

Science is a language spoken by every nation in the world. Put to a good use, science can show us how to feed millions of people. It can fight off diseases. It can transport people to the moon and beyond. Science is the language of the future.

The 21st century is almost here. By studying science, you are making a place for yourself in that future. Science creates more and more jobs every day. People are needed to run, fix, and program computers. They are needed to grow and protect our forests. And even more people are needed to work in hospitals, help keep the air clean, and grow food.

This book will teach you the main ideas of life, physical, and earth science. You will learn about the many different kinds of living things—from tiny bacteria to the great blue whale. You will learn how music travels through air. And you will learn what causes earthquakes. In the last chapter, you will learn about space exploration. When you finish this book you will be prepared to continue studying in any field of science that you choose. You will be ready to enter the 21st century.

Look for the notes in the margins of the pages. These friendly notes are there to make you stop and think. Sometimes they comment on the material you are learning. Sometimes they give examples. Sometimes they remind you of something you already know.

You will also find several study aids in the book. At the beginning of every chapter, you'll find **Learning Objectives**. Take a moment to study these goals. They will help you focus on the important points covered in the chapter. **Words to Know** will give you a look at some science vocabulary you'll find in your reading. And at the end of each chapter, a **summary** will give you a quick review of what you've just learned.

We hope you enjoy reading about science. Everyone who put this book together worked hard to make it interesting as well as useful. The rest is up to you. We wish you well in your studies. Our success is in your accomplishment.

What Is Science?

Unit 1

Chapter 1
The Wonder of Science:
From Atoms to Galaxies

Chapter 2
Science Skills: The Process of Discovery

Chapter 1

The Wonder of Science: From Atoms to Galaxies

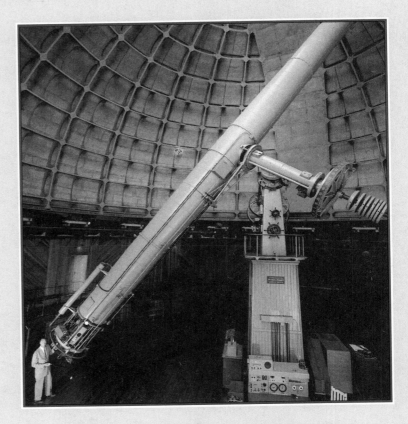

This scientist is looking at stars that are millions of miles away. How long do you think it takes for the light from these stars to reach the earth?

Chapter Learning Objectives
- Define science.
- Name two new technological discoveries.
- Name three different branches of science.
- Identify two careers in science.

Words to Know

atoms tiny parts into which all things on Earth can be broken down

experiments tests that are used to discover or prove something

field a profession or special area of interest

infinity space and time without end

observations careful studies of something, especially for scientific purposes

process a series of steps for making or doing something

science the study of nature and the universe, based on facts that are learned from observation and experiment

technology the application of scientific and industrial skills to practical use

universe everything that exists, including the Earth, sun, planets, stars and outer space

---

Imagine that you are outside on a clear night. You look up. What do you see? Stars, of course—thousands of them. But do you know what you are *really* seeing? You are seeing the stars as they were *many years ago!*

This is one of the amazing facts of **science**. The stars are so far away that it takes years for their light to reach Earth. In fact, some of the stars you see may not even exist anymore! They may have burned themselves out long ago. That's just a glimmer of how big and mysterious the universe really is.

The mysteries of science are often very tiny, too. Right at this moment, you are covered with different plants and animals. As you read, wildlife is snacking on your body! An eight-legged bug actually lives on your eyelids!

Science Is the Study of Our Universe

Atoms are tiny, tiny parts into which all things on Earth can be broken down. They are so tiny you cannot see them. Galaxies are huge groups of stars in the **universe**. Our own galaxy is called the Milky Way. It has more than 100 billion stars.

Science is our attempt to understand the universe. Scientists study everything, from the tiniest things on Earth to the **infinity** of outer space. In this book, you will learn how *you* fit into the science picture. You will learn how atoms as well as galaxies affect your life every day.

Science Changes Constantly

Many people think science is all fact. They think that a discovery is the truth forever. But science changes constantly.

People used to believe that the Earth was flat. They thought they had good reason to believe this. But of course now we know that the Earth is round. A new fact was discovered and an old fact became fiction.

One of the first people to do scientific research on plants and animals was Aristotle. He lived in Greece more than 2,000 years ago. Aristotle made a great many discoveries. But he also made some mistakes. For example, he observed that male horses have more teeth than female horses. So he decided that men must have more teeth than women!

The word "atom" means "that which cannot be divided." Scientists once thought the atom was the tiniest thing in the universe. Now they know the atom can be broken up into even smaller parts.

People used to be afraid to sail very far out to sea. They believed that they would fall off the edge of the world. What did Christopher Columbus and other explorers find out when they sailed across the oceans?

This does not mean Aristotle was a bad scientist. Science was still very young in those days. And scientists were just beginning to learn how to do research. For his time, Aristotle was a very good scientist. His work led to many discoveries by scientists who came much later.

Science Is Made of Facts and Processes
There are two parts to science. One part is the group of *facts*. These are the things that scientists have already discovered or proven. An example is the fact that plants need sunshine to grow.

The second part of science is the **process** of discovery. By reading this book, *you* are taking part in the process of discovery. The process is how people learn. It is the way one discovery leads to another. Scientists make **observations** and carry out **experiments**. This is the process by which scientists learn more science facts.

What people used to believe often seems very funny today. Perhaps in 2,000 years people will get their hands on this book. Some of our ideas may seem silly indeed.

On the Cutting Edge
Technology is science put to work. Every day new scientific discoveries are making life better for the world. Here are a few problems that have been solved very recently.

You already know many science facts. For example, how is the water in the ocean different from fresh water? Where does rain come from? What happens to snow on the ground when the sun shines on it?

- Record albums scratch easily. A worn-out needle, scratches, and dust get in the way of perfect sound. So scientists have come up with new technology to make listening to music even more wonderful. It uses a beam of light instead of a needle to play music. The beam of light is very narrow and very strong. It is produced by a device called a *laser*. The laser is built into a machine called a compact disk player. Music played on a compact disk is sharp and clear. Some people say it sounds like a live concert. And the compact disks never scratch or wear out.

- Braille is a system of writing for the blind. It was invented by the 19th-century Frenchman, Louis Braille, who himself was blind. Braille consists of a code of 63 characters. Each is made up of one to six raised dots. Many books have been translated into Braille. But the translation process is time-consuming and costly. Most books have not been translated. To solve this problem, a new machine has been invented. The Kurzweil Personal Reader uses a handheld scanner to read print electronically. A computer voice reads the words aloud. Now blind people can read their own letters, books, office papers, and anything else in print.

This is a picture of a Kurzweil machine. For fun, find out if your city library owns one. Try putting this book in it.

- One of the biggest food crops in Africa is called cassava. About 200,000,000 Africans depend on this root for food. In the early 1970s, a bug from South America began ruining these crops. If something isn't done, the crops will be completely ruined in the 1990s. But scientists think they have solved the problem. They have discovered another bug that eats the harmful one. In this way, scientists may have saved the food supply for millions of people.

Science Practice

Write answers to the following questions on a separate sheet of paper.

1. Name one example of technology in your home. What problem does that technology solve?

2. What are the two parts of science? Of which part are scientific experiments?

3. Name one scientific belief that people used to hold. What do people believe about that now? Do you think our beliefs will change? Why?

Take a Stand on Science

The new technologies in the examples on pages 6 and 7 all solved problems. But these answers were long in coming. Millions of dollars needed to be spent. Thousands of people had to make decisions. Science today is practiced by groups of people working together. Once in a while a single, brilliant scientist makes a big breakthrough. But usually, a lot of people help.

Here's where you come in. Even if you don't choose a career in science, you can help to make science decisions. Do you want your tax money to be spent on science to make more weapons? Science to make more food? Science for finding a cure for AIDS and cancer? Now more than ever, it is important for people to know something about science. As a voting citizen, you can help make these choices.

The Branches of Science

Science is a very large area of study. You will get a good overview in this book. You will study three branches of science: life science, physical science, and Earth science.

Life science is the study of living things and their parts and actions. Life scientists study plants, animals, humans, and other living things.

Physical science is the study of matter and energy. Matter is the stuff of which everything in the universe is made. Matter can be a solid, liquid, or gas. Physical scientists study matter. They study forms of energy such as light, heat, and electricity. And they also study how machines work.

Earth science is the study of Earth and its rocks, oceans, and weather. It is also the study of the sun, moon, planets, and stars.

Careers in Science

Science is one of the fastest growing career **fields** in the world. There are jobs in medicine, farming, computers, forestry, and building.

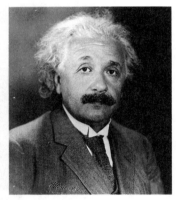

Albert Einstein was a scientist who became famous for his discoveries about matter and energy. Which branch of science did he work in?

People with science backgrounds work in zoos, factories, laboratories, hospitals, building sites, parks, and in kitchens. In fact, just about every field of work hires at least a few scientists.

Look at Appendix A at the back of the book for a list of many different science-related careers. Appendix B lists different institutions that hire people for science-related jobs.

People in Science: Walter Massey

In 1992, Walter Massey became the first African-American director of the National Science Foundation. NSF is an important federal agency that gives money for science research. Massey's job includes deciding what science projects the government should support. One important goal of his is to improve pre-college science education for women and minority students. If you were director of the National Science Foundation, what goals would you set for yourself?

Walter Massey

Chapter Review

Chapter Summary

- Science is the study of the universe. This includes everything from the tiniest things to the biggest things.
- There are two parts to science: a group of facts and a process of discovery.
- Both the group of facts and the methods of discovery change as science progresses.
- Technology is science put to work. Every day new discoveries lead to new technology.
- As voters, United States citizens can take part in making decisions about scientific work.
- The three main branches of science are life science, physical science, and Earth science.
- Each branch offers many career possibilities.

Chapter Quiz

Write answers to the following questions on a separate sheet of paper.

1. What is science?
2. What is technology?
3. Write at least three examples of technology that you can find in your home.
4. What is studied in life science?
5. What is studied in physical science?
6. What is studied in Earth science?
7. What scientific mistake did Aristotle make?
8. Name three decisions the government must make about how tax money should be spent on science.
9. Name three career fields in which there are science jobs.
10. The problem of the bugs and root crops in Africa is being solved by which branch of science?

Reporting on Science

Look at Appendix A at the back of the book. Choose the science career you would like best. Write the name of the career on a separate sheet of paper. Then write two reasons why you think you might like it.

Mad Scientist Challenge

Invent some new technology. On a separate sheet of paper, draw a picture or use words to describe a new machine. Write a sentence or two saying what the machine will do that is new.

Chapter 2

Science Skills: The Process of Discovery

Scientists study practically everything, from rocks to the way humans behave. Can you think of anything science cannot study?

Chapter Learning Objectives
- Name the five steps of the scientific method.
- Use a metric chart.
- List safety rules for science work in the laboratory.

Words to Know

gram a basic unit of weight in the metric system, equal to 0.035 ounces

liter a basic unit of volume in the metric system, equal to about 1.05 quarts of liquid

measurement the size, quantity, or amount of something

meter the standard unit of length in the metric system, equal to 39.4 inches, or slightly more than 3-1/4 feet

metric units the standard units of measurement in the metric system, based on the number ten and multiples of ten

unit a fixed amount or quantity that is used as a standard of measurement

---

What is the sun? If your answer is, "The sun is a star," you are correct. But, for that matter, what is a star? In ancient times, people worshipped the sun as a god. Somehow they knew that without the sun's light and heat, life on Earth would be impossible.

Thanks to the work of scientists, today we know that the sun is simply a star. And not a very big one at that. There are some stars that are one thousand times bigger than the sun! We also know that the stars are big hot balls of glowing gas.

Through the process of discovery, scientists have learned ways to measure a star's size and temperature. They can also find out how far away the star is from the Earth. None of this would have been possible without using the scientific method.

The Scientific Method

Scientists use a process of discovery called "the scientific method." There are five steps to this method.

1. **Say what the problem is**. There are an infinite number of problems to be solved. Take this one, for example: What is a cure for the common cold?

2. **Gather information**. First a scientist reads everything he or she can find on the subject. The scientist might talk to other scientists doing work on this problem. And he or she makes observations that might help to solve the problem.

3. **Suggest a good answer to the problem**. After a lot of research, the scientist suggests an answer to the problem. Let's say the scientist suggests that Vitamin C will cure a common cold.

4. **Test the suggested answer by running experiments**. Now the scientist must set up experiments to test the suggested answer. Twenty people with colds will be used. Ten will be given Vitamin C. Ten will not be given anything. This test may be run several times.

5. **Report the results**. Finally, the scientist will write up the results of the experiment.

Usually, the results of one experiment lead to another experiment. In the common cold experiment, ten people were given Vitamin C. Perhaps most of them got over their colds quicker than those who didn't take Vitamin C. But some who took Vitamin C did not get well. Why doesn't it work every time?

The scientist ran the experiment again. She paid special attention to those who took Vitamin C and didn't get better. She noticed that these people didn't sleep much during the experiment. The scientist then ran another experiment to test how sleep affects the common cold.

Science Practice

Write answers to the following questions on a separate sheet of paper.

1. What is the scientific method?

2. What is the first step in the scientific method?

3. After a scientist suggests a good answer to a problem, what is the next step?

4. Why does a scientist need to run experiments?

5. What is the final step in the scientific method?

6. Why do you think the scientist in the example above ran the experiment again?

7. What did the scientist decide to do after she ran the experiment a second time?

On the Cutting Edge

In May 1990, a severe storm dumped a shipment of sneakers into the Pacific Ocean. The 80,000 shoes went overboard about 800 kilometers southeast of the Alaskan Peninsula.

When Curtis C. Ebbesmeyer heard about the accident, he began combing the beaches for washed up sneakers. Was he looking for a free pair of hightops? Not exactly.

Ebbesmeyer is a scientist who studies ocean currents. He wanted to find out how ocean currents move through the Pacific Ocean. He knew this sneaker spill could help him gather important information.

Ebbesmeyer began calling beach bums, shell collectors, anyone who spent a lot of time on the shore. He asked people to collect washed up sneakers. He asked them to note exactly where and when the sneakers came to shore. Six months after the spill, sneakers began appearing on beaches in British Columbia, Washington, and Oregon.

Ebbesmeyer tracked down 1,300 shoes. Then, along with W. James Ingraham, Jr, he built a model showing the path of the drifting shoes.

At the end of 1992, the floating shoes were still turning up—now in Hawaii. The scientists are still following the floating footsteps. They predict that if they don't disintegrate, the shoes will eventually reach Asia and Japan.

The Metric System of Measurement

Many skills are needed to run science experiments. A very important one is **measurement**. Measurement is used to express the size, quantity, or amount of something. Your height and weight measure how much of you there is.

A **unit** is a fixed amount or quantity used by everyone when measuring. For example, when riding in a car, people talk about miles. A mile is a fixed unit. Everyone in this country knows what you are talking about when you say "a mile."

In most countries, however, people do not measure distance in miles. Neither do they use feet and inches to measure height, nor pounds to measure weight.

More than 90 percent of the world's people use metric units of measurement. **Metric units** are based on the number ten and multiples of ten, just like our money system. All scientists the world over use metric units. This makes it easier for them to talk about science with each other.

The United States is officially shifting from the use of inches, feet, and miles to the metric system. That's why highway signs sometimes give distance in miles and *in kilometers.*

Chapter Two   17

Thinking in metric units can be very confusing at first. But as you work with the metric system, you will get used to it.

Metric Units of Length

The main metric unit of length is the **meter**. A meter is equal to 39.4 inches. Most full-grown people are about 1-1/2 to 2 meters tall.

You know from money that "cent" means 1/100. A centimeter is 1/100 of a meter. "Milli" means 1/1000. A millimeter is 1/1000 of a meter.

A kilometer is one thousand meters. This is a little more than half a mile.

Metric Units of Area and Volume

Area is the number of square units needed to cover a surface. The area of this page is about 440.4 square centimeters. The area of the surface of the Earth is 510,100,000 square kilometers (196,951,000 square miles).

Volume is the amount of space an object takes up. The metric unit for measuring the volume of a liquid is called a **liter**. Perhaps you have bought soda in liter bottles. A liter is a little bigger than a quart. Often scientists work with very small amounts of liquids. They often use milliliters which are 1/1000 of a liter.

Metric Units of Weight

The metric system measures weight in units called **grams** and kilograms. There are about 28-1/3 grams in an ounce. A kilogram is equal to 1,000 grams or about 2.2 pounds. A metric ton is equal to 1,000 kilograms, or about 2,200 pounds.

Getting a Feel for Metric

- The thickness of a dime is about one millimeter.
- The width of your little finger is about one centimeter.
- The length of a golf club is about one meter.
- A distance of five city blocks is about one kilometer.
- The area of a button on a push-button telephone is about one square centimeter.
- An area of one square kilometer is large enough to hold about 180 football fields.
- The volume of a sugar cube is about one cubic centimeter.
- One liter is equal to about 2-3/4 cans of soft drink.
- The volume of a single bed is about one cubic meter.
- The weight of a United States dollar bill is one gram.
- The weight of one volume of an encyclopedia set is about one kilogram.
- The weight of a small automobile is about one metric ton.

Science Tools and Safety

Scientists work in laboratories, usually just called labs. For some scientists, a whole meadow might be their "lab." For others, a small room in a hospital is their lab.

The picture on page 21 shows several tools that scientists use. Any time scientists use these tools they are very careful. Working with fire and certain substances can be dangerous.

Science Alert

Here are a few of the safety rules that should *always* be followed.

1. Read every word in the instructions for an experiment before beginning. Before you start, be sure to have everything you need.
2. Keep a very clean work area. Do not have anything out that you do not need for the experiment.
3. Never taste or touch any substances in the lab unless instructed by your teacher or book to do so. Do not eat or drink or chew gum in the lab. Keep your hands away from your face.
4. Always make sure your equipment is in good condition. Never use chipped glass or frayed cords.
5. Whenever you are instructed to wear eye goggles, do so.
6. Know how to put out fires and where to find clean, running water.
7. Follow all instructions exactly.

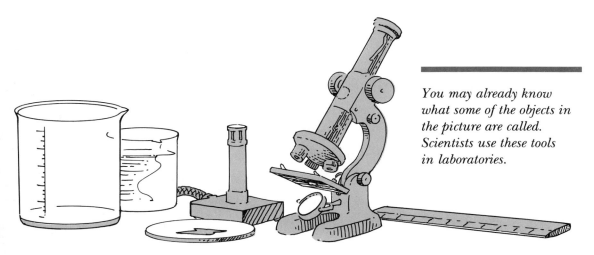

You may already know what some of the objects in the picture are called. Scientists use these tools in laboratories.

Beakers, petri dish, bunsen burner, microscope, metric ruler

A Lucky Mistake

Exact measurement and extreme cleanliness are very important in science labs. But everyone makes mistakes.

In 1928, Alexander Fleming was studying bacteria, which he was growing in small laboratory dishes. Some bacteria are harmful and cause sickness. Fleming wanted to discover a way to fight these bacteria.

One day he made a mistake on one of his experiments. He set the dish aside and forgot to throw it away. Some mold grew on the dish. Later, Fleming went to throw the dish away. But something made him stop and look at the mold more closely. He noticed that no bacteria grew near the mold. The mold was killing the bacteria!

This mold is called *penicillium*. Today it is made into penicillin, an important drug that fights bacteria. Fleming proved a point: the most important part of the scientific process is to *keep asking questions.* His "failure" became his greatest success.

Chapter Review

Chapter Summary

- There are five steps in the scientific method:
  1. Say what the problem is.
  2. Gather information.
  3. Suggest a good answer to the problem.
  4. Test the suggested answer by running experiments.
  5. Report the results.

- Often the results of an experiment lead to other experiments.

- Measurement is used to express the size, quantity, or amount of something. A unit is a fixed amount or quantity used by everyone when measuring.

- More than 90 percent of the world's people use metric units of measurement. Metric units are based on the number ten and multiples of ten, just like our money system.

- Safety in the lab is very important. Always read all the instructions before beginning an experiment.

Figuring Out Metric

Use words from the box to complete the following sentences. Write the sentences on a separate sheet of paper. Use the metric conversion chart in Appendix C if you need help.

more than	less than	the same as

1. An inch is _____ a centimeter.
2. A liter is _____ a quart.
3. A square kilometer is _____ a square mile.

Chapter Quiz

Write answers to the following questions on a separate sheet of paper.
1. What are the five steps in the scientific method?
2. Name at least two ways a scientist could gather information before suggesting an answer to a scientific problem.
3. What do the results of experiments usually lead to?
4. What is measurement?
5. What is a unit?
6. Why must scientists the world over use the same system of measurement?
7. What is the main metric unit for measuring length?
8. What is volume?
9. What is the metric unit for measuring the volume of a liquid?
10. What point did Alexander Fleming's mistake prove?

Mad Scientist Challenge

With which step of the scientific method did the sneakers help Ebbesmeyer? Explain how the sneakers helped him with this step

Unit 1 Review

Answer the following questions on a separate sheet of paper.

1. What are the two parts of science? Which one goes on in a laboratory?
2. Is Braille an example of technology? What problem did it solve?
3. Why did people once believe the Earth was flat? What kind of experiment finally proved the Earth was round?
4. How can you make a difference in the kinds of science that are funded by the government?
5. When you look at a star, chances are you are seeing how it was many years ago. Why is that?
6. Why do some speed limit signs show both miles and kilometers per hour?
7. How would you apply the scientific method to learn whether oil and water mix?
8. Why do you think scientists try experiments more than once?
9. It is important to make sure your laboratory equipment is in good shape. Why?
10. What is penicillin?

Life Science

Unit 2

Chapter 3
The Study of Life

Chapter 4
Living Things Are Made of Cells

Chapter 5
The Kingdoms of Life

Chapter 6
The Animal Kingdom

Chapter 7
The Plant Kingdom

Chapter 8
Genetics: The Code of Life

Chapter 9
Evolution

Chapter 10
The Human Body: Cells to Systems

Chapter 11
The Human Energy Systems

Chapter 12
Healthy Living

Chapter 13
Living Things Depend on Each Other

Chapter 3

The Study of Life

This bat is almost totally blind. Like all bats, it can fly with great speed and accuracy. Its built-in radar system prevents it from crashing into anything.

Chapter Learning Objectives
- Describe five very different life forms.
- Name five important fields in biology.
- List the five main characteristics of life.
- Give three examples of how an organism responds to its environment.

Words to Know

biology   the science that studies life

botany   the scientific study of plants

characteristics   the qualities or features of a person or thing

ecology   the study of how all living things on Earth depend on one another

environment   all the things around you

genetics   the study of how life characteristics are passed along to offspring

life span   the amount of time an organism is likely to live

microbiology   the study of organisms too small to be seen with the unaided eye

organism   a living thing

reproduce   to make more of one's kind

respond   to act in return or in answer to something

wastes   the leftover matter a cell or body does not need after it uses food for energy

zoology   the scientific study of animals

The Earth is about 93 million miles from the sun. This is *exactly* the right distance for us and all other life on Earth. Even if the Earth were just five percent closer to the sun, the heat would be unbearable. The oceans would dry up. Life would be impossible. What if the Earth were just one percent farther away from the sun? The whole planet would be completely covered with ice.

Luckily for us, the Earth is well set up for life. The temperature is good. There is plenty of water.

In this chapter you will learn about the qualities that all living things have. You will also learn why rocks, books, stereos, and fire are *not* alive.

Living Things Are Called Organisms

The Earth is swarming with life. Your school grounds alone are home for millions of plants, animals, and other living things. A living thing is called an **organism**. Life science, also called **biology**, is the study of all the organisms on Earth.

The variety of life forms is astounding. Even the differences in size are great. The adult blue whale weighs more than 90,000 kilograms (200,000 pounds). This is as much as about 23 elephants put together. The blue whale is about 27 meters (almost 30 yards) long. The giant redwood trees of California grow to a height of 90 meters (almost 300 feet). Their trunks can be more than four and a half meters (about 15 feet) thick. Yet the smallest organisms are so small you can't see them. Millions of them could swim in a drop of water.

The Fields of Life Science

Life science is a very big field of study. It is broken down into several smaller fields. Five important ones are botany, zoology, genetics, microbiology, and ecology. Most life scientists specialize in one of these areas.

Botany is the study of plant life. **Zoology** is the study of animal life. **Genetics** is the study of how life's characteristics are passed along to offspring. **Microbiology** is the study of organisms too small to be seen with the unaided eye. **Ecology** is the study of how all living things on Earth depend on one another. In this unit on life science, you will learn something about each of these fields.

Botanists study plant life.

On the Cutting Edge

Why only study life on Earth? What about life on other planets?

At the end of 1992, a group of scientists began a study to look for living beings on other planets. They are using very powerful radio telescopes, which pick up radio waves from space. Scientists hope that somewhere out there, intelligent beings are sending out messages that the radio telescopes may pick up. The project is called SETI—Search for Extraterrestrial Intelligence.

At first, people thought they were crazy, but today, most scientists agree that there must be life out there. Why would only one planet in the entire universe have life?

What Are Characteristics?

Characteristics are the qualities or features of a person or thing. Characteristics are what make one thing different from another. A few of your own characteristics include your hair color, height, shape, and the parts of your personality.

All living things share at least five important characteristics. These are the characteristics of all life. You share these characteristics with the blue whale, weeds, mushrooms—and every other organism on Earth.

Chapter Three   29

Every single thing, if it is alive, must: 1) get and use food; 2) move; 3) grow; 4) reproduce; 5) respond to its environment.

Science Practice
On a separate sheet of paper name two *characteristics* of each of these living things:

tree	fish	dog
flower	cow	worm

Getting and Using Food
How long do you think you would last without food and water? You might be able to live without food for a few weeks. You could not go without water for even half a week. All living things must have food and water.

Some animals hunt other animals for food. Others eat plants for food. Still others eat both animals and plants.

Once an animal has gotten food, its body must be able to use it. Body parts and chemicals work together to break the food down. After using the food for energy, the animal gets rid of the waste products.

Plants make their own food. They soak up energy from the sun. They also soak up water from the earth. Together, the sun energy and water combine to change into a kind of food. You will learn more about this later. Like animals, plants must also get rid of **wastes**.

Moving
Animals must be able to move for several reasons. They need to find food. They must be able to move

Lions need meat in order to survive.

away from danger. And they also need to be able to find mates. Different animals move in very different ways. Some fly, others walk and run, and still others swim.

Plants also move, though it is harder to notice. Plants need to get sunlight.

Try This Experiment
Put a green plant in a room with one window. Within a day or two, the plant's leaves will be facing the window. After several days, the whole plant will probably lean toward the window. Plants always move in the direction of light. Remember that plants use sunlight to make their own food.

Growing
All living things grow. After a certain size is reached, organisms stop growing. People stop growing taller when they are between 15 and 20 years old. Some trees grow for hundreds of years before reaching their full height.

Reproducing
All organisms can **reproduce** themselves. This means they are able to make more of their own kind of organism. When a cat has kittens, she has reproduced. When humans have babies, they have reproduced.

Plants reproduce in many different ways. Many reproduce using seeds. The seeds must fall or be carried to a good place to grow. Wind, water, and animals often help plant reproduction by scattering seeds.

Science Practice

Use the picture below to answer the following questions. Name two organisms for each question. Write your answers on a separate sheet of paper.
1. Which animals are moving away from danger?
2. Which animals are moving to get food?
3. Which plants have the easiest time getting sunlight?
4. Which animals walk?
5. Which animals fly?
6. Which animals swim?

Responding to the Environment

Everything that surrounds you makes up your **environment**. If you are in the classroom, your environment includes the other students. It also includes desks and chairs, paper, a chalkboard, and the air in the room. All the things you can see, hear, or feel around you are parts of your environment.

To **respond** means to act in return or in answer to something. If a friend pokes your arm with a pencil, you will respond. You may tell him or her to cut it out. You may grab your arm in pain. You may do any number of things. But you will be *responding*.

All living things respond to their environments. Zebras respond to the sight or smell of lions by running for safety. Garden plants respond to deep watering by growing deep roots.

How Long Do Organisms Live?

All living things die at some time. However, different organisms have different life spans. A **life span** is the number of years a type of organism is likely to live. A mayfly lives its entire adult life in one day. Most dogs live for about 14 years. The white pine lives for almost 500 years. The bristlecone pine lives almost 5,000 years.

In the United States the average human life span is now about 73 years. Earlier this century, the average life span was about 65 years. Progress in medical science and nutrition has helped to lengthen the human life span.

Some scientists think the average human life span can increase for only so long. Others hope that new discoveries will help them find a way to slow down aging. Perhaps by the year 2200, people will live to be several hundred years old.

What is the world's largest organism? Scientists believe it may be a giant underground fungus discovered in the woods under the Wisconsin-Michigan border in April of 1992. It has taken over 37 acres and weighs between 100 and 1,000 tons, and it is still growing!

Chapter Review

Chapter Summary

- Living things are called organisms. Five important fields of biology are botany, zoology, genetics, microbiology, and ecology.

- All organisms share five important characteristics. They get and use food, move, grow, reproduce, and respond to their environments.

- Animals must hunt their food, be it other animals or plants. Plants are able to make their own food by using sunlight. All organisms must get rid of wastes after using food for energy.

- Animals move to find food, mates, or to get away from danger. Plants move to get closer to sunlight. All organisms grow for a certain amount of time and then stop growing.

Chapter Quiz

Write answers to the following questions on a separate sheet of paper.
1. What name is used to describe all living things?
2. What is another name for life science?
3. What do we call scientists who specialize in studying plants?
4. What do we call scientists who specialize in studying animals?
5. Give three reasons animals must be able to move.
6. Give one reason plants must be able to move.
7. What does reproduce mean?
8. Name ten things in your immediate environment.
9. How might a lion respond to the smell of fresh meat?
10. How might a plant respond to a dark room?

Reporting on Science

Do you think there is life on other planets? On a separate sheet of paper write two reasons why or why not. Use your imagination.

Mad Scientist Challenge

How many times longer is the blue whale than you? Divide your height into the length of the blue whale. The length of the blue whale can be found on page 28. Write your answer on a separate sheet of paper. If you want, try the same thing with your weight.

Chapter 4

Living Things Are Made of Cells

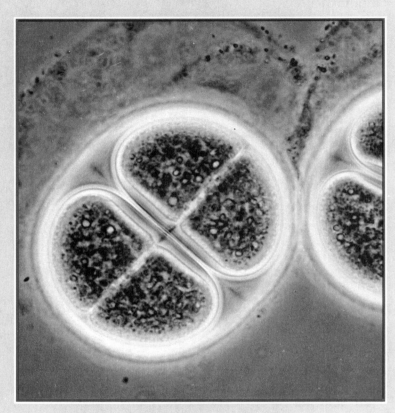

All living things are made of cells.

Chapter Learning Objectives
- Describe cells and their parts.
- Contrast plant and animal cells.
- Describe how cells get and use energy.

Words to Know

cell the tiny basic unit of which all living things are made

chloroplasts the special parts of a plant cell that help make food from sunlight

cytoplasm the watery substance in a cell

digest to break down food inside the body or cell into a form that can be used

DNA molecules in the nuclei of cells that control many of the characteristics of living things

elements the basic substances of which all matter is made

membrane the protective covering that holds a cell together

microscope a machine for viewing objects that are too small to be seen by the naked eye

molecule two or more atoms joined together

nucleus (plural, nuclei) the part of the cell that controls all the other parts

respiration the way a cell gets energy by mixing food and oxygen

vacuoles openings in a cell that store food, water, or wastes

The first **microscope** was invented nearly 400 years ago. With microscopes scientists could study objects too tiny to see with the naked eye. This opened up a whole new world to scientific investigation.

In 1665 scientist Robert Hooke used the microscope to look at a piece of tree bark.

Through the microscope, the bark looked like it was divided into many boxes. Hooke thought the boxes looked like the small rooms monks lived in. These rooms are called cells. So Hooke called the boxes he saw in the tree bark "cells," too.

Robert Hooke also studied many other plants under the microscope. They all seemed to have these cells. Hooke's discovery proved to be very important. We now know that the **cell** is the basic unit of plant and animal life.

Three new elements have been discovered in the last 25 years. However, these elements have not been officially approved or named.

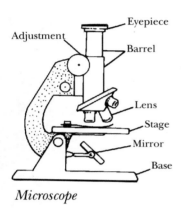

Microscope

Matter Is Made of Elements

Cells are matter. Remember that matter is anything that takes up space. Everything, living or nonliving, is matter. All matter on Earth can be broken down into one or more kinds of elements. There are 103 **elements**. Oxygen, carbon, helium, and hydrogen are a few examples of elements. The table of chemical elements in Appendix D lists all the elements on Earth.

Elements can be broken down into even smaller parts. These tiny particles are called *atoms.*

Atoms are so small that scientists can only see them with special, very strong microscopes.

Different kinds of atoms often join together to form the materials in our world. Look at the diagram on this page. It shows the kinds of atoms (or elements) found in your body.

When two or more atoms join together, they form a **molecule**. Water molecules are made when two hydrogen atoms join one oxygen atom.

People in Science: Wilbert and Bob Gore

Water molecules are bigger than air molecules. Knowing this, Wilbert Gore and his son Bob had an idea. Why not make cloth with a weave too tight to let in water molecules but loose enough to let *out* air molecules? The cloth would be waterproof. But it would also "breathe," meaning let hot air out. If you have ever worn old-fashioned rubber rain gear, you'll know what a good idea this was!

It took them many years, but the Gores finally made just such a cloth. Today Goretex, their cloth, is made into tents, jackets, hats, and all kinds of things that need to be waterproof and also "breathe."

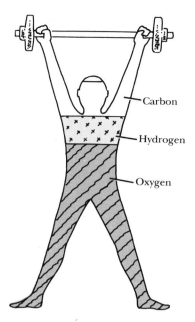

The human body is 65 percent oxygen, 18 percent carbon, and 10 percent hydrogen. The rest of the body is made of small amounts of nitrogen, calcium, phosphorus, and other elements.

All Living Things Are Made of Cells

Everything in the universe is made of atoms. But only living things have cells. Cells are the building blocks of life. Think of how bricks are put together to make a house. In this same way, cells are put together to make a living thing.

A cell is something like a very tiny water balloon. It is runny inside and has a thin outer covering. Dogs, ants, flowers, and even people are all made of cells. So is athlete's foot, an organism called a *fungus*, that grows on people's feet.

Though you cannot see cells without a microscope, they are much bigger than atoms and molecules. Many molecules go into making a cell.

Your own body has about 100 trillion cells. This is twenty thousand times greater than the number of people in the world. But there are many organisms made of just one cell. These organisms can only be seen with a microscope.

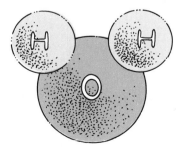

A water molecule is made of two hydrogen atoms and one oxygen atom.

Chapter Four 39

The skin is the outer covering of the body. It protects the organs inside the body from injury or disease. Which part of a cell is most like your skin?

The Main Parts of a Cell

Most of a cell is made up of a watery, sometimes gooey, substance called **cytoplasm**. The other parts of the cell float around in this watery cytoplasm.

All cells are surrounded by a **membrane**. This thin covering holds the cell together. It lets food pass into the cell. The membrane also lets wastes pass out.

Near the middle of the cell is the nucleus. The **nucleus** is the cell's "command post." It controls all other parts of the cell.

The big open spaces are called **vacuoles.** These work as storerooms for the cell. Cells use vacuoles to store food, water, and wastes.

Science Practice

Take out a sheet of paper. Copy this picture of a cell. Write the names of all the parts next to them in your picture.

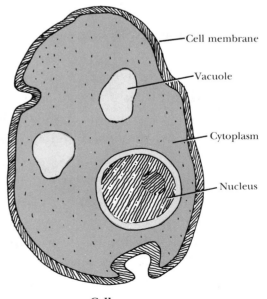

Cell

Answer the following questions on a separate sheet of paper.
1. How did the invention of the microscope help scientists?
2. About how many elements are there?
3. What holds the cell together and lets food pass into it?
4. Can an organism be made of just one cell?
5. What special information about molecules did the Gores use to make their cloth?

Amazing Science Fact
Sponges are a kind of animal that live on the ocean floor. They are one of the simplest and oldest kinds of animal alive today. Yet they can do some biological tricks that even people cannot do.

If a person loses an arm or a leg, he or she does not get a new one. Human cells will heal over the open area. But we cannot grow new arms or legs.

Sponges, however, can grow new parts. If part of a sponge is cut off, the cells reproduce to replace that part.

Some scientists think humans could once do that. They believe that somewhere along the line we have lost that biological trick. In this one way people have been outdone by the simple sponge.

How Cells Get Energy

Each cell in your body has all the characteristics of life. Cells can get and use food. They can move and grow. Cells can also reproduce and respond to their environments. In order to do these things, cells need energy.

Cells get their energy from food and oxygen. Food passes through the cell membrane into the cytoplasm. Oxygen also passes through the membrane into the cell. The food is broken down, or **digested**, in the cytoplasm. This digested food mixes with the oxygen.

When digested food and oxygen mix, energy is released. This is called respiration. **Respiration** is the release of energy from the breakdown of food in a cell.

Besides energy, respiration also produces certain wastes. Cell wastes are products that are not needed by the cell. They are leftovers. These wastes leave the cell through the membrane.

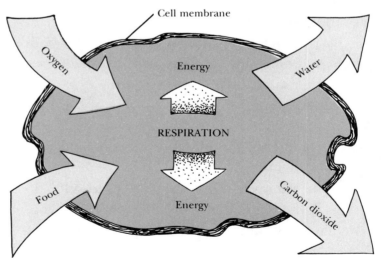

FOOD + OXYGEN = ENERGY + WASTES

The wastes from respiration are water and carbon dioxide.

Differences Between Plant and Animal Cells

Plant and animal cells have a few important differences. First, plant cells have cell walls. The cell wall is outside the membrane. It is harder and stronger than the membrane.

Plant cells usually have bigger vacuoles than animal cells. This is because plant cells must store a lot of water. Often, however, animal cells have more vacuoles than plant cells.

The most important difference is that plant cells have something called **chloroplasts**. Chloroplasts store a green coloring. This green coloring traps sunlight. Plants use the trapped sunlight to make their own food.

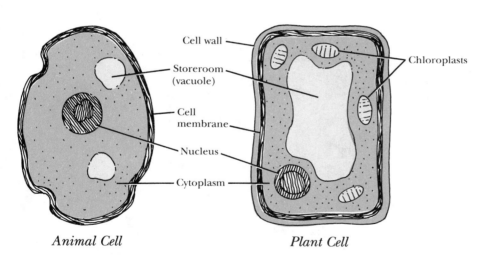

Animal Cell   *Plant Cell*

The Importance of DNA

Inside the nucleus of all cells is a very important kind of molecule called **DNA**. DNA is one of the largest molecules found in living things. It is made of thousands of smaller molecules joined together.

These smaller molecules are joined together in a certain order. The order of the molecules forms a "life code." This life code controls all activities of the cell.

DNA controls how cells grow and multiply. It controls whether an organism will grow into an anteater, a human, or another kind of life. DNA controls whether a person will have brown or blue eyes. It controls whether a gorilla will be tall or short. You will learn a lot more about DNA in Chapter 8.

DNA molecule

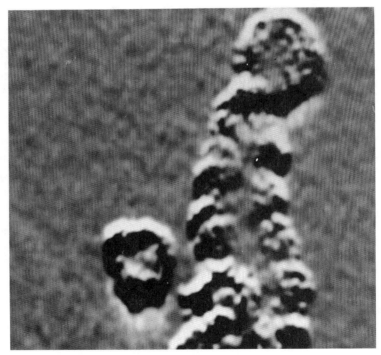

The first photograph ever taken of DNA

On The Cutting Edge

In 1953 James Watson and Francis Crick devised a model of the structure of DNA. The model was shaped like a twisted ladder. Based on research, the model explained how cells are able to reproduce themselves. In 1963 the two scientists were awarded the Nobel Prize. Many people consider the Nobel Prize to be the highest honor a scientist can get. In 1989 the first photograph ever taken of DNA appeared in the newspapers. The photograph proved Watson and Crick's theory about the structure of DNA.

Today, knowledge about DNA is being used in criminal courts. Since every person has a different DNA code, scientists can identify people by the DNA in their blood. Sometimes, a person's guilt or innocence can be told with this DNA information.

Chapter Review

Chapter Summary

- With the invention of the microscope came the discovery of cells. All living things are made of cells.

- All matter, living and nonliving, is made of atoms. When atoms join together they make molecules. Many, many atoms and molecules go into making a cell.

- Most of a cell is made up of a watery substance called cytoplasm. A membrane surrounds the cell and holds it together. The membrane also lets food pass in and wastes pass out.

- The nucleus of a cell controls all the other parts. The vacuoles store water, food, and wastes.

- Cells get energy by respiration. This is the way that a cell takes in food and oxygen through its membrane. When the food and oxygen mix, energy and wastes are released.

- Plant cells differ from animal cells in three important ways. Plant cells have cell walls. Plant cells have bigger, but usually fewer, vacuoles. Most important of all, plant cells have chloroplasts. Chloroplasts hold a green coloring that helps plants make food from sunlight.

- DNA is an important molecule (or group of molecules) found in the cell nucleus. The exact way the molecules are arranged controls the characteristics of an organism.

Chapter Quiz

Answer the following questions on a separate sheet of paper.

1. Who first discovered cells in tree bark?
2. Give three examples of elements.
3. What is formed when two or more atoms join together?
4. What are the building blocks of life?
5. What is the watery, sometimes gooey, part of the cell?
6. Which part of the cell is the "command post"?
7. What three things can be stored in vacuoles?
8. What two things must go into a cell for it to get energy by respiration?
9. What is the most important difference between plant and animal cells?
10. Which group of molecules form a "life code" in the nucleus of cells?

Reporting on Science

Copy this drawing of two cells on a separate piece of paper. Decide which is a plant cell and which is an animal cell. Write "plant" under one and "animal" under the other. Below each picture explain why it is a plant or animal cell. Use the following words in your explanation.

one large vacuole	no chloroplasts
chloroplasts	cell wall
several smaller vacuoles	no cell wall

Chapter 5

The Kingdoms of Life

Every organism on Earth belongs to one of the five kingdoms of life.

Chapter Learning Objectives
- Explain why a classification system of organisms is important.
- Name the five kingdoms of life and give an example of each.

Words to Know

algae (singular, alga) plant-like protists.

bacteria (singular, bacterium) simple, one-celled organisms that are visible only through a microscope; a kind of moneran

biologist a scientist who studies the habits and growth of living organisms

classification the way in which biologists group organisms by type

fungi (singular, fungus) organisms that cannot move around like animals and do not have chloroplasts like plants. Mushrooms, molds, and yeasts are fungi.

kingdoms the five main groups in biological classification

monera (singular, moneran) tiny organisms that have some nucleic materials, but no true nuclei, in their cells, such as bacteria

protists tiny one-celled organisms that are neither plants nor animals but that often have characteristics of both

protozoans (singular, protozoan) animal-like protists

species the smallest groups in biological classification

structure the way an organism is put together

vaccine a substance injected into a person to keep him or her from getting a certain disease

What is another name for a mountain lion? Did you answer cougar? Puma? Panther? All of these are different names for the same kind of cat.

People often call the same plants and animals by different names. These are the common names of organisms.

But think about what would happen if *scientists* used these common names. They would confuse each other very quickly. More than a million different kinds of animals are known today. And at least 324,000 different kinds of plants are known. **Biologists** the world over use the same name for each organism. That way they can communicate with each other about their studies. In this chapter, you will learn more about how scientists group organisms for study.

THE FIVE KINGDOMS

Kingdom	Description	Examples
1. Moneran	One-celled Have no true nuclei	Bacteria Blue-green algae
2. Protist	One-celled Have nuclei	Protozoans Algae
3. Fungus	Multi-celled Have cell walls Cannot make food	Yeast Molds Mushrooms
4. Plant	Multi-celled Have cell walls Use sunlight and chlorophyll to make food	Seed plants Evergreens Ferns
5. Animal	Multi-celled Cannot make their own food Eat other organisms	Insects Reptiles Birds Mammals

Classification of Organisms

Grouping organisms by type is called **classification**. Biologists study the characteristics of organisms to decide how they should be grouped. Overall, there are five big groups. These are called **kingdoms**. The chart shows the five kingdoms of life.

You already know two of the kingdoms: the plant and animal kingdoms. The biggest difference between these kingdoms is the way the organisms get their food. Animals must go out and hunt or grow their food. Plants make their own food by using chloroplasts in their cells and sunshine.

Science Practice

Study the above picture. Then make two lists on a separate sheet of paper. Head one list "Plant Kingdom." Head the other list "Animal Kingdom." Under each list write as many examples as you can find in the picture.

Smaller Groups of Organisms

In each of the five kingdoms there are smaller classifications of organisms. For example, the animal kingdom is broken into two important groups. Animals with backbones are in one group. Animals without backbones are in the other. And these two groups are broken into even smaller groups.

A **species** is the smallest classification group. A species includes all the organisms that reproduce together. Humans, for example, are a species of primates.

You will read much more about plants and animals in the next two chapters. In this chapter, you will read about the other three kingdoms.

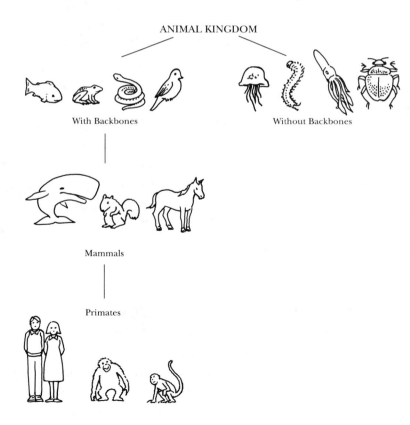

People in Science: Aristotle

Remember Aristotle, that mighty thinker from ancient Greece? He formed one of the first classification systems. He classified animals by where they lived. He closely observed all the animals he could find. He noticed that they all lived in one of three places: in the air, in the water, or on land.

Aristotle's system was a good start. But there were still many problems. For example, where do penguins fit into this picture? They live in the water and do not fly. Yet they have wings instead of flippers. Today penguins are classified as birds.

Whales and dolphins point out another problem in Aristotle's system. They live in the water and they certainly look like fish. But neither whales nor dolphins have scales like fish. More important, though, is the fact that they breathe air just like land animals. On closer study, whales and dolphins are more like humans and tigers than fish.

Scientists now classify organisms by their structure. An organism's **structure** is the way its body is put together.

Even though they cannot fly, penguins are classified as birds.

The Protist Kingdom

For hundreds of years, people believed there were only two kingdoms, plant and animal. Then the microscope was invented. It showed new kinds of organisms. Today we call them protists.

Protists are one-celled organisms. Many of them have both plant and animal characteristics. That's why biologists decided they must be in their own kingdom, the protist kingdom.

Some protists are plant-like. These are called **algae**. You may have seen algae growing on lakes or floating in the sea. It is usually green, red, or brown. What you see is really a giant group of algae. Most individual algae are too small to see without a microscope.

Protozoans are animal-like protists. They often have tiny shells. When they die, these shells pile up on the ocean floor. The chalk in your classroom is made from the buildup of these shells.

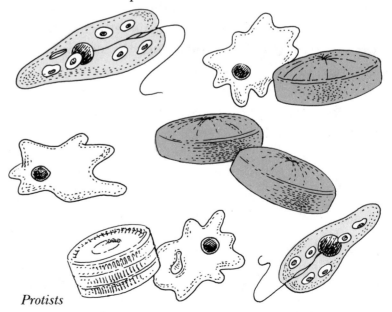

Protists

On the Cutting Edge

Most protozoans are harmless. But some kinds cause a serious disease called malaria. This disease strikes 270 million people each year, killing more than 2.5 million. Scientists may have found a way to fight this killer protozoan. They have developed a vaccine that works on mice with malaria. However, the malaria that mice get is a little different from the malaria that people get. Scientists must now use this information to make a vaccine that works on people.

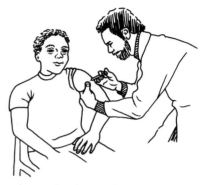

A vaccine is a substance injected into the body that prevents you from getting a certain disease.

The Moneran Kingdom

You may have heard of **bacteria**. Bacteria are microscopic organisms. They are everywhere.

They live in the ocean, the sky, and even in your skin. Many bacteria are harmlessly living inside your body right now.

Some bacteria are very useful. We use them to make many kinds of foods, such as cheese and yogurt. But, other kinds of bacteria are very dangerous. They can cause sickness and even death.

Scientists discovered something very strange about bacteria. Unlike animals, plants, or protists, they do not have *true nuclei* in their cells. But they do have DNA and other nucleic materials. These nucleic materials aren't all held together in one place. They float around in the cytoplasm. For this reason, scientists decided they must create another kingdom: the moneran kingdom.

Blue-green algae are another kind of **monera**. While most algae are protists, blue-green algae do not have true nuclei.

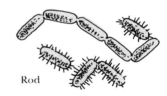
Rod

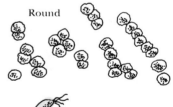
Round

Spiral

The three shapes of bacteria

Science Alert

A few kinds of bacteria are very dangerous. One type (called *Clostridium botulinum*) produces a deadly poison. It has been said that only fifteen ounces of this poison could kill everyone in the world! People can get this poison from foods that were not canned in the right way. That's why it's not a good idea to buy food in cans that are crushed or open.

The Fungus Kingdom

Have you ever left bread out of the refrigerator for a long time? Something soft and fuzzy grew on it, right? This fuzzy substance is called *mold*. Mold is part of the fungus kingdom.

People used to group **fungi** with plants. After all, they have a lot in common with plants. They grow in one place. And they do not move around looking for food as animals do. But fungi are missing one important plant characteristic: chloroplasts. That means they cannot make their own food. Fungi can't be monera either because they *do* have true nuclei in their cells. Finally, fungi are too big to be protists. So scientists made the fungus kingdom classification just for these organisms.

The fungus kingdom is made up of molds, yeasts, and mushrooms. Some of these fungi get their food from dead organic matter. Others eat the living organisms on which they live. For example, athlete's foot is a fungus that eats the skin on people's feet.

The kind of mushrooms you buy in the store are good to eat. But some wild mushrooms are deadly poison. Never eat mushrooms you find in the wild, unless you're with an expert.

Amazing Science Fact

You have already learned that cells mix food and oxygen to get energy. This process is called respiration. The respiration of yeast cells is very helpful to bakers.

Bakers put yeast in bread dough. These one-celled organisms "eat" the bread as it cooks.

Remember that respiration produces wastes. One of these wastes is carbon dioxide. This is the gas that causes bubbles in soda. When yeast produces carbon dioxide as a waste, it makes gas bubbles in the baking bread. This causes the bread to rise.

Chapter Review

Chapter Summary

- Scientists must use the same names for organisms in order to communicate with each other. Classification is a way of grouping organisms by their type. Organisms are grouped into five kingdoms: animal, plant, protist, moneran, and fungus.

- Each of the kingdoms is broken down into many more groups. The smallest groups of classification are called species. A species includes all the organisms that can reproduce together.

- Protists are organisms that have some plant and some animal characteristics. Algae are plant-like protists. Protozoa are animal-like protists.

- Monera are organisms that do not have true nuclei. They do have some nucleic material that floats around in the cytoplasm. Bacteria and blue-green algae are the two kinds of organisms in the moneran kingdom.

- Fungi are organisms that cannot move around like animals. Nor do they have chloroplasts for making their own food. Molds, yeasts, and mushrooms are all members of the fungus kingdom.

Chapter Quiz

Write answers to the following questions on a separate sheet of paper.

1. Name the five kingdoms and give one characteristic of each.
2. What has to be true for a group of organisms to belong to the same species?
3. What species are you a part of?
4. Name the two big groups the animal kingdom is divided into.
5. Why are penguins not in the same group as fish?
6. Why did biologists decide that protists needed to be put into their own kingdom?
7. Why were bacteria put in their own kingdom?
8. Which fungus is helpful in making bread?
9. How are fungi different from plants?
10. Why must organisms be classified?

Mad Scientist Challenge

Choose one of the following two topics. Look it up in an encyclopedia. On a separate sheet of paper write two interesting facts that you find on the topic.

1. **bubonic plague:** the sickness that killed millions in Europe in the 1300s
2. **botulism:** the deadly poison people can get from eating foods that have not been properly canned

Chapter 6

The Animal Kingdom

The lion has been called the king of the jungle.

Chapter Learning Objectives
- List the three characteristics common to all animals.
- Name three invertebrates.
- Name three vertebrates.
- Describe what makes an animal a mammal.

Words to Know

amphibian an organism that lives part of its life in water and part on land, such as a frog

appendage a body part that sticks out, such as a wing, feeler, arm, or leg

arthropod an animal with a shell, jointed appendages, and body divided into segments

cold-blooded having a body temperature that changes with the temperature of the environment

gills organs that allow fish to get oxygen from water

host an animal on or in which parasites live

invertebrates animals without backbones

mammals animals that have hair on their bodies and give birth to live young. Female mammals provide milk for their young.

migrate to move at regular times of year from one region to another for food, mating, or warmer temperatures

parasite an organism that lives on or in another organism

vertebrates animals with backbones

warm-blooded having a body temperature that stays the same in both hot and cold weather

You already know a lot about the human animal. After all, you are one. In this chapter you will learn some surprising facts about other animals in our biological kingdom. For example, humans have 656 muscles. That sounds like a lot, right? Well caterpillars have about 2,000!

What Makes an Organism an Animal?

The protist, moneran, and fungus kingdoms are made up of very simple organisms. Humans are far more complicated. But humans have many things in common with these organisms. For one thing, we share the five characteristics of life (review page 29). We are also made of cells, just like protozoa, algae, bacteria, and mushrooms.

But there are many differences between humans and these simple organisms. Three important characteristics make a human an animal and not a plant, protist, moneran, or fungus.

- Animals can move around at will. Plants move their leaves and bend their stems toward sunlight. But they cannot choose to get up and go to live in a different part of the forest. Animals can fly, swim, run, or walk.

- Animals must go out and get their own food. Plants can make food using chloroplasts in their cells. Animals must find other plants or animals to eat.

- Finally, animals are made of specialized cells. Plants also have specialized cells. But protists, monera, and fungi do not.

What Are Specialized Cells?

Most of the organisms you have read about so far are one-celled. A few, like mushrooms, are made of many cells. But all the cells in a mushroom are pretty much alike.

Animals have many different kinds of cells. Every animal cell has a nucleus, a membrane, vacuoles, and the other cell parts. But different animal cells have different shapes and sizes depending on what jobs they do. The cells of animals are *specialized*. That is, each cell has a certain job.

Think of a one-celled organism as a company run by one person. That person has to do all the work. But bigger companies hire many workers. In big companies, people have specialized jobs. Some people type, others keep the books, and still others build or sell things. The bigger the company, the more specialized jobs it has. This is also true of the cells in animals. The more complicated the animal, the more specialized cells it has.

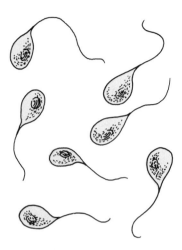

Sperm cells have tails. They swim to meet egg cells.

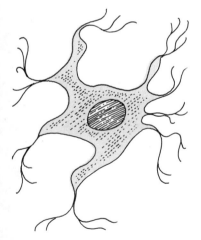

Nerve cells are long and string-like. They carry messages around the body.

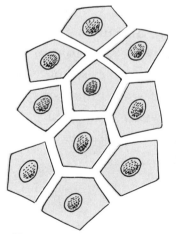

Skin cells are flat and broad. They protect the organism.

Muscle cells must be able to stretch.

Chapter Six

The Invertebrates

Invertebrates are animals that do not have backbones. They are much simpler creatures than the **vertebrates**, animals that do have backbones.

Some examples of invertebrates are sponges, jellyfish, worms, insects, and shellfish. In this book, you will study just a few invertebrates.

Sponges

Sponges have been around longer than any other animals. They came into being more than a billion years ago. They grow on rocks on the ocean floor. Sponges get their food by filtering the water. As the ocean washes past and through them, the sponge cells make use of food bits in the water.

The colored, evenly-shaped sponges in your kitchen today are probably human-made. But perhaps you have seen a natural sponge. It probably was a tan color and had an uneven shape. Natural sponges are really the skeletons of living sponges.

Sometimes sponges have been mistaken for plants. They spend most of their lives fastened to one rock. However, when they are young, they move around like animals. Also, they catch their food as animals do.

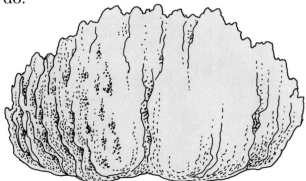

Sponges have changed little in the last billion years.

Worms

Worms are another type of invertebrate animal. There are many kinds of worms. The tapeworm is a ribbon-like flatworm. Some tapeworms are as long as 18 meters (about 44 feet). The tapeworm is a **parasite**. That means it lives inside another animal called a **host**. A host is any animal that supports a parasite. The tapeworm absorbs the host's digested food.

Roundworms have more parts than tapeworms. Roundworms are able to digest their own food. They have a mouth and an *anus*. The anus is the opening through which wastes leave the body.

The earthworm is the kind of worm people dig up for fishing. It has even more parts than the roundworm. Earthworms have a *crop* for storing food and a *gizzard* for grinding food. They also have five hearts for pumping blood to the ends of their bodies.

Earthworm and tapeworm

Mollusks

The word "mollusk" means "soft-body." Snails, slugs, clams, oysters, squid, and octopuses are all mollusks. Although most live in water, some snails and slugs live on land. Most mollusks have hard shells that protect their bodies.

Snails are useful in aquariums. They eat the algae that build up on the aquarium glass. This keeps the glass clean and easy to see through.

Spiny-Skinned Animals

Sand dollars and starfish belong to another group of animals called *spiny-skinned*. They live in salt water.

Oyster farmers hate starfish because they eat oysters. A starfish uses its five strong arms to pull the oysters open. Then the starfish actually pushes its own stomach out of its mouth. It puts its stomach into the oyster shell and digests the oyster. Then it swallows its own stomach again.

Starfish are spiny-skinned animals.

Arthropods

Spiders, scorpions, cockroaches, ticks, crabs, lobsters, bees, mosquitoes, ants, and grasshoppers are all **arthropods**.

Most arthropods have shells of some kind. They all have jointed **appendages**, or parts that stick out of the body. Wings, legs, arms, and feelers are all appendages. Because they are jointed these appendages can bend. The bodies of arthropods are also divided into segments.

There are more than 800,000 species of arthropods. Arthropods are broken into many groups called "classes." The number of body segments and number of appendages decide which class an arthropod is in. In this chapter you will study three classes: insects, crustaceans, and spiders.

Amazing Science Fact

Grasshoppers can jump over things 500 times their own length. If you could do that, you would be able to jump over a tall city building.

Honeybees must fly about 50,000 miles to gather enough nectar to make one pound of honey.

Insects

Insects are the largest group of arthropods. Their bodies are divided into three parts: head, thorax, and abdomen. Insects have three pairs of legs and usually two pairs of wings. They also have feelers, called *antennae*. Insects are the most plentiful and widespread of all land animals.

Science Alert

Many insects are useful to people. But some insects do people a lot of harm. For example, fleas can get diseases from rats and give those diseases to humans. Flies and cockroaches bring germs from garbage and animal waste into people's homes.

Crustaceans

Lobsters, crabs, and crayfish are examples of crustaceans. Crustaceans have two body segments and five pairs of legs. The front pair are usually called claws. The other four pairs are walking legs. Crustaceans are useful to people as food.

Spiders

Many people think spiders are insects. But spiders have only two body segments instead of three. They have four pairs of legs. And they do not have antennae. Scorpions, mites, and ticks all belong in the same classification as spiders.

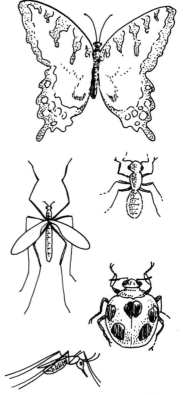

Butterfly, crane fly, ant, ladybug, mosquito

Science Alert

The black widow spider is very dangerous—and not only to people. A female black widow bites and kills a male spider right after mating.

Black widow

Science Practice

For each statement, write *true* or *false* on a separate sheet of paper.
1. All animal cells do the same job.
2. Invertebrates are animals without backbones.
3. Sponges are plants.
4. Starfish are helpful to oyster farmers.
5. Arthropods have shells, jointed appendages, and segmented bodies.
6. Lobsters, crabs, and crayfish are all crustaceans.

The Vertebrates

Vertebrates are animals with backbones. Humans, zebras, birds, fishes, rats, dogs, horses, and frogs are all vertebrates.

Vertebrates have much more complicated bodies than the invertebrates. They have developed brains, digestive parts, and reproductive parts, among other things. And most vertebrates also have appendages.

There are five main groups of vertebrates: fish, **amphibians**, reptiles, birds, and mammals.

Fish

Fish are the simplest vertebrates. Salmon, trout, sharks, and minnows are all fish. They are **cold-blooded** animals. That means that their body temperature changes with the temperature of their environment. If the water is cold, the fish will be cold. If the water is warm, the fish will be warm.

Fish have fins and scales. Look at the picture of a scale on this page. You can see that it has rings on it.

The whale shark is the largest fish in the world. It can grow up to 39 feet in length. But it feeds on the tiniest plants and animals in the sea.

The scales on a fish grow with the fish. Each year one more ring is added to each scale. You can tell how old a fish is by counting the rings on its scales.

Fish do not have lungs to breathe air. They get their oxygen right out of the water. They use **gills** to do this. As water goes through the gills, oxygen is absorbed right into the blood. The blood carries the oxygen to all parts of the fish's body.

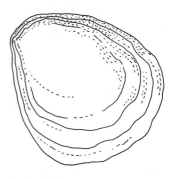

Look at this fish scale. How old is the fish it belongs to?

Amphibians

Salamanders and frogs are two kinds of amphibians. They have wet, slippery skin and two pairs of legs. Most amphibians have sticky tongues which they use to catch insects. Like fish, they are cold-blooded.

The word "amphibian" means "able to live on both land and in water." Frogs, for example, spend most of their lives on land. But they lay their eggs in water. Amphibian eggs have no shells. They would dry out if left on land.

Young amphibians have gills like fish. They use their gills to get oxygen from the water. When they get older they leave the water to live on land. By then they have developed lungs for breathing air. But most amphibians must still live in wet places. They need to keep their skin wet. When a frog is completely underwater, it breathes through its skin.

People in Science: Katherine Milton

The Mayoruna Indians of Brazil's Amazon taught Katherine Milton, a scientist from California, something very important about frogs. Milton watched the Mayoruna Indians collect the slime

Katherine Milton

that covers frogs' bodies. They dried the slime on sticks. Then they burned the skin on their own arms. Finally, the Mayoruna Indians put the dried frog slime into their wounds.

At first, the men got very sick and fell asleep. But when they awoke, they were able to hunt for long hours and not get tired. The Mayoruna men "take frog" at least once a month. The Mayoruna women "take frog" when they need to work long hours.

This "secret magic" may be a major medical discovery. Researchers have found that frog slime is an amazing healer. They believe it may lead to breakthroughs in fighting strokes, Alzheimer's disease, depression, and seizures.

Katharine Milton is not surprised. She studies the native people of the Brazil Amazon because she knows they have great knowledge. Modern scientists do not always respect the knowledge of people who have lived close to nature for centuries. As far as Milton is concerned, the Mayoruna Indians are the teachers and she is the student.

Reptiles

Many people say that snakes are slimy. But if you were to touch a snake, it would feel warm and dry. From time to time, a snake sheds the outer layer of its dry, scaly skin.

Reptiles are cold-blooded land animals. Some examples are lizards, turtles, snakes, alligators, and crocodiles. They have lungs for breathing air. Most reptiles have four legs and clawed toes. Snakes are an exception. Because they have no legs, they move along on their bellies. Some snakes have as many as 300 pairs of ribs.

Reptiles are able to live in very dry places. Their bodies are covered with a set of hard scales, almost like plates. This lets a lizard bake in the hot sun without drying out. The hard coating keeps in its moisture.

Reptiles lay leathery eggs. The leathery cover protects the young reptile.

Amazing Science Fact

Crocodiles swallow their food whole. A full-grown crocodile carries about five pounds of rock in its stomach. The rock helps grind up the food so the crocodile can digest it.

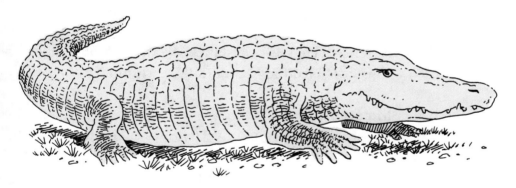

Birds

There are more than 8,600 different kinds of birds. Hummingbirds are the smallest. A hummingbird weighs about one tenth of an ounce. Ostriches are the largest birds. An ostrich can weigh as much as 300 pounds.

Birds have feathers and wings for flying. They also have hollow bones which make them lighter. The lighter they are the easier it is for them to stay in the air. Birds also have two legs.

Amazing Science Fact
Some birds are very fast. A gannet can dive through the air at speeds of more than 100 miles per hour. The peregrine falcon is believed to be able to dive at 180 miles per hour.

Birds are **warm-blooded**. That means they keep a constant body temperature, no matter what the environment is like.

Many birds **migrate**. That means they fly long distances each year to reach warmth and better feeding grounds. Sometimes they migrate to nest or mate. The blackpoll warbler, a bird about the size of a sparrow, flies from Canada to New England each year. It stays in New England for about two weeks eating insects. Then it flies—nonstop—to South America. That is a distance of about 2,500 miles. This little bird flies day and night without resting.

Mammals
You are in the group of vertebrates called **mammals**. Humans, elephants, mice, whales, tigers, bears, and squirrels are all examples of mammals.

Mammals are warm-blooded animals that have hair on their bodies. The females' bodies make milk to feed their young. Mammals also give birth to live young. This means that the babies are not born inside eggs. At birth, they are already formed as young animals.

The ancestors of whales lived on land and had four legs. But they moved into the sea and their bodies gradually changed. Their front legs developed into flippers. Two tiny hipbones are the only remaining traces of their hind legs.

Mammals also have big brains. They have two sets of appendages and many have tails. Most mammals live on land. But whales, porpoises, and dolphins are three kinds of mammals that live in the water. The blue whale is the biggest mammal on Earth. The shrew is the smallest mammal.

Did You Know?
You may have heard of a *flock* of sheep or a *herd* of cows. Flock and herd are group names. Take a look at these other group names.

Animals	Group Name	Animals	Group Name
Frogs	Army	Chicks	Brood
Badgers	Cete	Quail	Covey
Geese	Gaggle	Lions	Pride
Gorillas	Band	Ants	Colony
Fish	School	Wolves	Pack

Chapter Review

Chapter Summary

- Animals are different than plants, protists, monera, and fungi in three ways: 1) Animals can move around at will. 2) Animals must get their own food. 3) Animals are made of specialized cells.

- All animals are divided into two main categories: vertebrates and invertebrates. Vertebrates have backbones. Invertebrates do not have backbones.

- Sponges, jellyfish, worms, insects and shellfish are examples of invertebrates.

- Humans, zebras, birds, fish, rats, dogs, horses, and frogs are all examples of vertebrates.

- The five main groups of vertebrates are: fish, amphibians, reptiles, birds, and mammals.

- Vertebrates are much more complicated than invertebrates. They have developed brains, digestive parts, and reproductive parts. Most vertebrates also have appendages.

- Mammals are animals that have hair on their bodies and give birth to live young. Female mammals produce milk for their young.

Chapter Quiz

Answer the following questions on a separate sheet of paper.

1. Name three ways in which animals are different from plants.
2. Name three kinds of specialized cells in animals.
3. What are invertebrates? Give three examples.
4. Give three examples of mollusks.
5. Name the three segments of an insect's body.
6. How are spiders different from insects? Give at least three differences.
7. Why isn't Katherine Milton surprised that frog slime is a great healer?
8. Name two cold-blooded animals and two warm-blooded animals.
9. What kind of bones do birds have that make flying easier?
10. Most mammals have tails. Name one kind of mammal that doesn't.

Make a Vertebrate Chart

On a separate sheet of paper, make a chart like the one below. Use the following words to fill it in.

fish	hair or fur	2 legs, wings
amphibians	wet, slippery skin	4 legs
reptiles	feathers	fins
birds	wet scales	2 to 4 legs
mammals	dry scales	usually tails

Group	Body Covering	Appendages

Chapter 7

The Plant Kingdom

We rely on rain forests like this one to produce the air we breathe.

Chapter Learning Objectives
- Describe the appearance and function of plant roots, stems, leaves, flowers, seeds, and fruits.
- Explain the process of photosynthesis.
- Identify the reproductive parts of a flower.

Words to Know

carbon dioxide a gas made of carbon and oxygen molecules

chlorophyll the green coloring in chloroplasts that traps sunlight

germination the process by which a young plant breaks out of its seed

petals the colored outer parts of a flower that help protect its inner parts and that attract insects

photosynthesis the process by which plants make sugar using sunlight, water, chlorophyll, and carbon dioxide

pistil the female part of a flower

plant ovary the fruit of a plant that grows around the seed

pollen the powdery dust on the ends of stamens that holds plant sperm cells

pollination the transfer of pollen from the stamen to the pistil of a plant

sperm male sex cells

stamens the male parts of a flower

We owe a lot to plants. Without them, humans and all other animals could not survive. We would have no food to eat and no air to breathe. In this chapter, you will find out why animal life would be impossible if there were no plants.

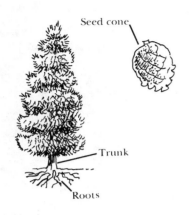

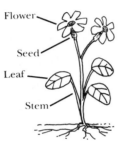

Different plants need different amounts of water. A desert plant's roots would rot if you gave it the same amount of water as a forest plant. What would happen to the roots of forest ferns if they were planted in the desert?

What Are Plants?

Plants are organisms made of many specialized cells. They are different from animals in three basic ways. 1) They make their own food. 2) Their cells have cell walls. 3) They cannot move from place to place at will. But, like all other living things, plants need oxygen, water, and food.

There are many kinds of plants. Most of the plants you see around you are *seed plants*. Those are the kind you will study in this chapter. Seed plants have roots, stems, leaves, and seeds.

Some seed plants have flowers and produce seeds inside fruits. Avocados, daisies, and plum trees are examples of flowering seed plants. Another kind of plant produces seeds in cones. Pine trees, for example, produce their seeds in pine cones.

Roots Get Water from Soil

When you pull up weeds, you pull on the top of the plant. The parts of the plant below the surface of the soil are called roots. Sugar beets and carrots are examples of *roots*.

Roots have several jobs. For one thing, they hold the plant in place. Roots also store extra food that the plant has made. But their most important job is to soak up water and minerals from the soil. Water enters a plant through the roots. Then it moves to all parts of the plant that need it.

The Stem Is the Pathway for Food and Water

Stems are the upright parts of plants. They run from the ground up to the leaves. The stems of trees are called *trunks*.

Stems hold a plant up so its leaves can get sunlight. Stems also carry water and food. First the roots soak up water. Then the water is delivered up the stem to all other parts of the plant.

The leaves make food for the plant. And that food is delivered *down* the stem to the roots.

You have probably eaten plant stems. Asparagus, rhubarb, and sugar cane are all stems. An unusual stem is the potato. Some people think the potato is a root. But a potato is really a part of the stem where the plant is storing extra food underground.

Try This Experiment
Fill three glasses with water. Put a different color food coloring in each one. Then put a carnation in each glass. Wait one day. Explain what this experiment shows about the job of stems.

Leaves Are the Food Makers
You already learned that plants make their own food. They use a green coloring stored in chloroplasts in the plant cells. This green coloring is called **chlorophyll**. Most of a plant's chlorophyll is found in its leaves.

The way in which plants make food is called **photosynthesis**. Four things are needed for photosynthesis: chlorophyll, sunlight, water, and carbon dioxide. **Carbon dioxide** is a gas in the air we breathe. It is made of carbon and oxygen molecules. Here's how photosynthesis works.

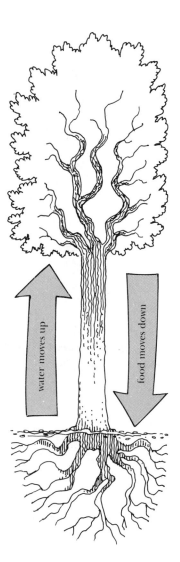

Water moves up the trunk of a tree. Food moves down the trunk.

The parts of a leaf. Water + Chlorophyll + Sunlight + Carbon Dioxide = Sugar + Oxygen.

Water comes from the roots through the stem to the leaves. Carbon dioxide from the air enters into small openings under the leaves. As sunlight strikes the chlorophyll in the leaves, energy is released. This energy makes the carbon dioxide and water join together to make sugar.

This is not the kind of sugar you buy at the store. This is a very simple sugar. It is the basis of all food. Animals eat the plants and get the energy stored in this sugar. Then people eat the animals or the plants and also get the energy. So you can see that *all* animals depend on photosynthesis for food.

Animals get something else they need from photosynthesis: oxygen. When the sugar is made, there is some leftover oxygen. This goes out of the plant through small openings on the undersides of leaves. Without plants covering the Earth and releasing oxygen, we could not breathe.

Science Alert

Rain forests grow in very warm, wet places. They produce great amounts of oxygen. Also, rain forests provide food for millions of animals.

In Brazil, people are cutting down the thick plant growth in rain forests to create farmland.

Biologists all over the world are concerned about the loss of the rain forests. Many biologists believe that the soil of the rain forests cannot support farm crops. The biologists believe that instead of gaining lush farmland, Brazil will end up with desert. The loss of the rain forests may change the environment of the entire world.

Science Practice

Think of all the ways you use plants in your life. Write your answers to the following questions on a separate sheet of paper.

1. Name ten different plants you eat.
2. Name one plant that can be used in building homes.
3. Name one plant that is used to make fabric for clothing.
4. Do you know of any plants that are used as medicines? Name them.

Flowers Are the Seed Makers

Flowers hold the reproductive parts of seed plants. The picture on this page shows the petals, pistil, and stamens of a flower.

The **petals** are outer parts of the flower. They are usually brightly colored. The **pistil** is the female part of the flower. At the bottom of the pistil is the ovary. The **plant ovary** is the fruit of a plant that grows around the seed. It is where the female sex cells, called eggs, are stored.

The male parts of the flower are called **stamens**. At the end of each stamen is a light powdery dust called **pollen**. Pollen holds the male sex cells called **sperm**.

For seed plants to reproduce, pollen must land on a pistil. Sometimes wind blows pollen off the stamens. But, there is so much pollen that it often lands on the pistil of another plant. The pistil is

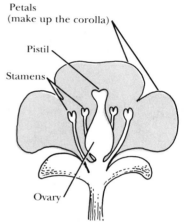

The parts of a flower

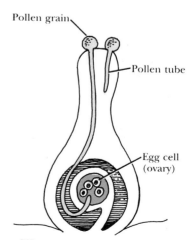

The process of fertilization

You can probably guess that bees don't help just to be kind. They are after something called nectar in the flower. They use the nectar to make honey.

sticky and traps the pollen. This process is called **pollination**.

Bees often help with pollination. They are drawn to the bright colors and nice smells of flowers. The bees have sticky legs. So when they land on a stamen, the pollen sticks to their legs. They fly off. When they land on another flower, they leave a little pollen on the pistil.

After pollination, a sperm cell goes down the pistil to the ovary. There it joins with an egg cell. Then a seed begins to grow.

Seeds Are Young Plants in a Protective Coat

You may have eaten sunflower seeds. Seeds are also found in apples, squashes, and other foods.

Seeds are tiny new plants in a protective coating. In other words, seeds hold the young offspring of plants. Most of the seed is food for the young plant. It has a lot of growing to do and needs a lot of food. Often seeds wait weeks, even years, before growing into a plant. A seed will wait until the environment is warm and wet enough. Then it will grow into a plant.

Wind and water carry seeds far from the plants that made them. Sometimes animals carry seeds, too. Some kinds of seeds stick to the coats of animals. An animal may carry such a seed a great distance before the seed falls off. Of course, some seeds never make it to a good place for growing.

But if a seed gets a chance, it will grow. It will break through the hard seed coat that has been protecting it. This is called **germination**.

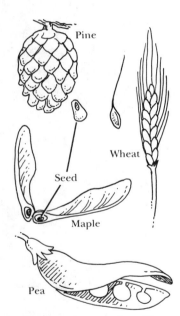

Which of these seeds will travel well on the wind? Which will roll along the ground? Which will travel on the fur of an animal?

Fruits Are the Ovaries of Plants

Split open a watermelon. You will see hundreds of black seeds. All the rest of the melon is protecting those seeds.

The fruit is really just the ovary of a plant. The ovary gets bigger when a seed begins forming in it. The ovary, or fruit, protects the seed from water loss, disease, and insects.

Many of the foods we often call vegetables are really fruits. Anything that has seeds in it, but is not a cone, is a fruit. This includes tomatoes, beans, avocados, eggplants, even oats and wheat.

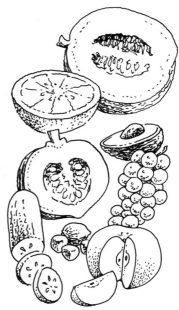

The fruits protect their plants' seeds.

Amazing Science Fact

There are a few plants that actually eat meat! These plants grow in soil that lacks some nutrients they need. So these plants catch insects to get the nutrients.

Pitcher plants have leaf parts that are shaped like pitchers. This part is lined with brush-like hairs. The hairs point down into the pitcher. So when an insect happens along, it slides right in and cannot get out. The plant digests the insect.

The Venus flytrap also catches insects on its brush-like leaves. The leaves close like a trap around the insect.

Why, then, are these classified as plants rather than animals? Because they are rooted to the ground. They cannot move around at will. They are also called plants because they still get most of their food from photosynthesis.

Chapter Review

Chapter Summary

- Plants are organisms with specialized cells. They make their own food. They cannot move from place to place.
- The roots of a plant soak up water and minerals for the plant. They also hold the plant in place.
- The stem holds the plant upright. It is also a pathway for food and water. Water from the roots travels up the stem. Food made by the leaves travels down the stem.
- Leaves carry out most of a plant's food-making. This process is called photosynthesis.
- The flower of a plant is where the seed is made. Stamens, which hold pollen, are the male parts of the flower. The pistil, which has the ovary at its base, is the female part of the flower. Pollination takes place when pollen lands on the pistil.
- Seeds are young plants in a protective coating. They can live for weeks, sometimes years, before growing into plants. They must be in a wet, warm place to germinate.
- Fruits are the ovaries of plants. The ovary of a flower swells after the egg cell has been fertilized. The fruit protects the seeds from disease, water loss, and animals.

Chapter Quiz

Write answers to the following questions on a separate sheet of paper.
1. Name four parts common to all seed plants.
2. Name three things roots do for plants.
3. What does a stem carry up a plant?
4. What does a stem carry down a plant?
5. What is the green coloring in chloroplasts called?
6. What four things are needed for photosynthesis?
7. What are the products of photosynthesis?
8. What is the female part of a flower called?
9. What are the male parts of flowers called?
10. How do plants help you breathe?

Scientific Drawing

Look at the picture of the flower. Copy the picture on a separate sheet of paper. Draw an arrow that points to each part. Next to the arrow, write the name of the part. Use the following words:

| roots | pistil | stem | stamen |
| petals | leaf | ovary | pollen |

Reporting on Science

Take a notebook and pencil to a grocery store. Make three headings on your paper: Stems, Leaves, Roots. Find as many foods as you can to go under each heading. Be sure to look at the canned foods as well as the fresh.

Stems	Leaves	Roots

Chapter Seven

Chapter 8

Genetics: The Code of Life

No two people are ever exactly alike. Even identical twins are different from each other. Can you think of any reasons why?

Chapter Learning Objectives
- Name five of your own traits.
- Describe how traits are passed from one generation to the next.
- Define genes and chromosomes.
- Explain how genetics can help farmers improve their crops.

Words to Know

breeding   producing offspring; raising plants or animals, especially to get new or better kinds

chromosomes   thread-like parts of a cell nucleus made up of DNA and genes

crossbreeding   combining the sex cells of organisms with different traits to create new traits

fertilization   the joining of a sperm cell with an egg cell

gene   a part of a chromosome that controls the development of individual traits

heredity   the passing of traits from parents to offspring

hybrids   the offspring of crossbreeding

mutation   a change in the genetic code of an organism

traits   characteristics, which may be inherited, that identify organisms as individuals

Have you ever wondered why you look the way you do? Or why you look like some members of your family but not like others? Which of your features look like your mother's or father's? Your nose? Your body size and shape? The color of your skin and eyes? Do you talk or act like any of your sisters or brothers?

In what ways do you look *different* from other members of your family? Do you have any features that none of them have?

All the ways you look and act are called **traits.** The color of your skin, hair, and eyes, and your size and shape are all traits. You also have personality traits. Are you quiet or talkative? Are you careful or reckless? These are different personality traits.

In this chapter you will learn some of the reasons why you have the traits you do.

What Is Heredity?

All living things reproduce. This means they produce new organisms to take their place when they die. Humans, for instance, have babies. Plants make seeds that grow into more plants. These new organisms are called *offspring*.

Offspring get many of their traits directly from their parents. The passing down of traits from parents to offspring is called **heredity**. The scientific study of heredity is called *genetics*.

The science of genetics is very young compared to other sciences. It began just a little more than 100 years ago. Yet today, genetics is one of the fastest growing sciences. There are new breakthroughs every year.

The Beginning of Genetics

The study of genetics started in the mid-1800s with an Austrian monk named Gregor Johann Mendel. In the garden at his monastery, Mendel grew 22 different kinds of pea plants. He was **breeding** these plants to study how traits were passed from parents to offspring.

Some of Mendel's pea plants were short and bushy. Others were tall and climbing. Some had white

Mendel was ahead of his time. He first announced his discovery in 1866. But his results were ignored until the beginning of this century.

flowers, others had pink flowers. Some produced round seeds, others wrinkled seeds.

Mendel took the pollen from a tall plant. He used this pollen to pollinate a short plant. This is called **crossbreeding**. He also crossbred white-flowered plants with pink-flowered plants. Then he crossbred round-seeded plants with wrinkled-seeded plants.

When plants with different traits are crossbred, the offspring are called **hybrids**.

Science Practice

On a separate sheet of paper, try to guess what came of Mendel's experiments.
1. In crossbreeding tall and short plants, did he get a) tall plants b) short plants c) medium-height plants?
2. In crossbreeding white-flowered plants with pink-flowered plants, did he get a) white-flowered plants b) pink-flowered plants c) light pink-flowered plants?
3. In crossbreeding round-seeded plants with wrinkled-seeded plants, did he get a) round-seeded plants b) wrinkled-seeded plants c) partly wrinkled-seeded plants?

Mendel's results may surprise you. When he crossed tall and short plants he always got tall plants. When he crossed pink- and white-flowered plants, he always got pink-flowered plants. When he crossed round-seeded plants with wrinkled-seeded plants, he always got round seeds.

Gregor Mendel

	Dominant	Recessive
seed shape	round	wrinkled
seed color	yellow	green
seed coat color	white	colored
pod shape	full	pinched
pod color	green	yellow
flower position	side	end
stem length	long	short

The seven characteristics that Mendel studied in pea plants.

Chromosomes, genes, and DNA molecules are so small they all fit in the nucleus of a single cell. Yet the amount of information they hold is mind boggling. Studying genetics is almost like studying a whole other universe.

Mendel had discovered that some traits were stronger than others. He found that a strong trait from one parent may cover or hide a weaker trait from the other. In pea plants, tallness was a stronger trait than shortness. The color pink was a stronger trait than white. And round seeds were stronger than wrinkled ones.

The stronger traits are called *dominant traits*. The weaker traits are called *recessive traits*. Recessive traits are not lost. They may show up in later generations. That explains how two brown-eyed parents can have a blue-eyed child. Both parents carried a recessive trait for blue eyes. They both passed on that trait to the child.

Chromosomes and Genes

Scientists now know that traits are controlled by DNA, a special kind of molecule found only in the nuclei of cells. DNA is arranged in thread-like structures called **chromosomes**.

Each chromosome holds thousands of pieces of information about an organism's traits. Each of these bits of information is called a **gene**. Genes are the basic building blocks of heredity.

All chromosomes come in pairs. Half are from the father. Half are from the mother. That means all genes come in pairs, too.

Every inherited trait is controlled by at least one pair of genes. Most traits are controlled by more than one pair of genes.

Each species has a certain number of chromosomes. Humans have 23 pairs. That means that every nucleus in every human cell has 46 chromosomes. Fruit flies have only 8 (4 pairs).

How an Organism Gets Its Genes

As you know, all living things are made of cells. Most cells reproduce by division—a single cell divides into two cells. As the cell is doing this, it copies its own nucleus exactly. That way both new cells are complete. They are also exactly the same.

Many organisms, including humans, begin with two special kinds of cells. These are the sex cells. They are called sperm and eggs. Like body cells, the sex cells reproduce by dividing. But they are different in one important way. They do not copy all the chromosomes in their nucleus. Each new cell gets only half the chromosomes.

Then, a sperm and an egg cell join. The two cells become one cell with a full set of chromosomes. This is called **fertilization**. The offspring that result will have genes from both the sperm cell and the egg cell.

Scientists often use fruit flies to study genetics. Their chromosomes are large and easy to see. They have hundreds of offspring at a time. Also, fruit flies reproduce every ten days.

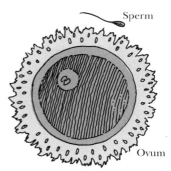

Sperm cell and egg cell

Mutations

New traits may suddenly appear in organisms. For example, a purple petunia plant may have a red flower. After that, some of the offspring may have the same red flowers. Or a cat may be born without a tail. Some of its offspring may also be born without a tail.

This happens because of a change in a gene or a chromosome. Something has changed the "genetic code" of the DNA. Maybe, during fertilization, the chromosomes of sperm and egg cells hooked up just a little differently. This change in the genetic code of an organism is called a **mutation**.

Chapter Eight 91

Do you think this mutation would help this snake to survive and reproduce?

Mutations, once they occur, are passed along to offspring. But only if the mutation does not kill the organism first. Many mutations do cause harm or death. Imagine a mutation that causes a bird to be born with a soft beak. The bird would not be able to crack nuts or dig for worms. It would die.

Once in a long while, though, a helpful mutation occurs. Imagine a mutation that causes a giraffe to have a longer neck. It could reach more leaves on a tree than other giraffes. Its offspring would have an easier time than shorter-necked giraffes.

On the Cutting Edge

Diabetes is a disease that keeps the body from using sugar properly. Until recently, diabetes was a disabling and often fatal disease. Then, in 1922, researchers found that people with diabetes could be helped by a substance called insulin. But insulin was very expensive and hard to get.

Now, however, scientists can remove insulin from DNA molecules. They attach the insulin to bacterial DNA. Because bacteria reproduce very quickly, the insulin in the bacteria gets reproduced quickly, too. Today bacteria "factories" make lots of insulin quickly and inexpensively. Many diabetics who would have died without insulin can now lead normal lives.

Plant and Animal Breeding

Long before anyone knew anything about DNA, people used genetics. Breeding is the matching of organisms for reproduction. Farmers do this to control the traits of the offspring of their plants and animals. Even in ancient times farmers bred the best fruit trees to get better fruit. They bred the best milk cows to get better milk.

Today scientists, farmers, and gardeners carefully choose plants and animals for breeding. This is called selective breeding.

Some people think scientists should not interfere with genetic codes. Why do you think they feel that way?

Environment and Traits

Your genes do not control everything about you. Your environment also plays a big part in the formation of your traits.

Suppose, for instance, that a person has the genes to be a very fast runner. But suppose she has a poor diet and never exercises. These environmental factors would probably keep her from being very fast.

A tomato plant may be the offspring of parents that produced big juicy tomatoes. The young plant has very good genes for producing delicious fruit. But suppose the plant is rooted in poor soil. And it is not watered enough. Despite its good genes, the tomato plant may not produce any fruit at all.

Chapter Review

Chapter Summary

- All the ways you look and act are called traits. Many traits are passed from parents to offspring. This is called heredity. Genetics is the study of heredity.
- Mendel used pea plants to study genetics. He discovered that some traits are dominant and that others are recessive.
- Chromosomes are found, always in pairs, in the nuclei of cells. Genes are parts of chromosomes. Each gene determines some trait in the organism.
- All cells reproduce by dividing. Body cells copy their nuclei before dividing. Sperm and egg cells do not copy their chromosomes before dividing. Offspring get half their chromosomes from one parent and half from the other. When the sperm and egg cell join, it is called fertilization.
- A change in the genetic code is called a mutation.
- People have been using select breeding for a long time to make better plants and animals.
- Environment also plays an important part in the development of traits.

Chapter Quiz

Write answers to the following questions on a separate sheet of paper.

1. Which family member do you look most like? List three traits the two of you share.
2. What did Mendel get when he crossbred tall and short pea plants?
3. What two kinds of traits did Mendel discover?
4. DNA is arranged in thread-like structures called what?
5. What are the basic units of heredity called?
6. How many pairs of chromosomes do humans have?
7. What is fertilization?
8. What is a change in the genetic code called?
9. Give two examples of select breeding.
10. How can the environment help form an organism's traits? Give two examples.

Describing Traits

Choose any plant or animal. The organism may be at your home or near school. Choose a pet, if you wish. On a separate sheet of paper name as many of its traits as you can. Be sure to include its size, shape, and color. Also describe how it feels to your touch and how it smells. Describe how the organism acts. Name three traits that are caused by its environment. Name three traits that are caused by its genes.

Mad Scientist Challenge

Think of two plants or animals you think would make an interesting hybrid. Use your imagination. On a separate sheet of paper draw a picture of the new plant or animal you have made by crossbreeding.

Chapter 9

Evolution

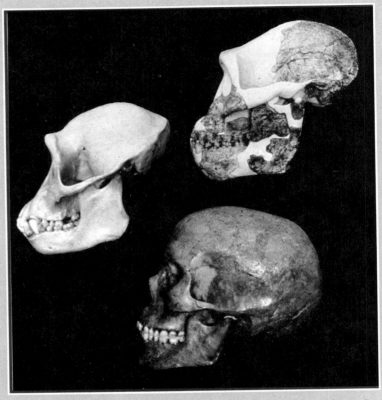

One of these skulls belonged to a human. One of them belonged to a chimpanzee. And one of them belonged to a human-like animal that lived several million years ago. Can you tell which is which? Do you see any similarities?

Chapter Learning Objectives
- Identify two important processes of evolutionary change.
- Explain the difference between natural selection and mutation.
- Name three things that scientists study to learn about evolution.

Words to Know

evolution changes in a species over time

evolve to change over time

extinct no longer living on the Earth; referring to a species that has died out

fossils the remains of organisms that lived long ago

natural selection the way that those organisms best suited to their environments survive and pass their helpful traits along to offspring

paleontology the scientific study of fossils

theory a group of ideas or principles that explain why something happens

---

Sometime, at least 3 billion years ago, living organisms first appeared on Earth. They were tiny one-celled creatures, a lot like protists. They lived in the ocean.

All the plants and animals alive today **evolved** from those first creatures. Even humans! But human-like animals have only been around for about 5,000,000 years. That's really not very long. Not if you consider that the Earth has existed for about 4,600,000,000 years.

In this chapter, you will learn about some of the processes of **evolution**. From those first one-celled creatures, these processes formed all the organisms alive today.

Evolution Is Change Over Time

Species change over time. For example, the woolly mammoth no longer exists. But that species slowly changed into what we know today as the elephant. Dinosaurs don't exist anymore either. But many scientists think that today's birds evolved from dinosaurs. This process of change in species, usually over thousands or millions of years, is called evolution.

Scientists have put together the science of evolution from many clues. For instance the woolly mammoth has been **extinct** for a long time. But scientists found a woolly mammoth frozen in Siberia. Ice had preserved it for 25,000 years. The bones and teeth of dinosaurs and other extinct animals have also been found. Extinct insects have been found trapped in hardened tree sap. These are the kinds of things scientists use to construct their picture of evolution.

Scientists found a woolly mammoth that had been frozen for 25,000 years.

The remains of organisms are called **fossils**. Fossils show that organisms from the past are different from organisms on Earth today. They also show that today's organisms evolved from these older species.

The scientific study of fossils is called **paleontology**. Studying fossils is a lot like detective work. Paleontologists put together a long story from a few clues. By studying fossils, scientists have learned that the first horses were about the size of large modern dogs. Early horses had four toes instead of one. Their teeth were shorter than the teeth of today's horses.

Fossils are remains or imprints of plants or animals that lived a long time ago.

Early horses were the size of large modern dogs.

Try This Experiment

Collect two leaves that are alike. Place one in a jar with a lid. Put small holes in the lid. Place the other leaf in a clear, plastic container. Fill the container with water and put it in the freezer. Compare the conditions of the two leaves after a couple weeks. Explain why glaciers can be helpful to scientists studying fossils.

On the Cutting Edge

Most scientists think the dinosaurs disappeared from Earth very quickly. No one is sure why. Some think a large meteorite crashed into Earth. They say the meteorite caused great clouds of dust when it fell. Because the clouds of dust blocked the sun, the weather on Earth suddenly changed. Many forms of life died.

In 1991, scientists found strong, new evidence for this argument. They found a crater, 112 miles (180 km) wide, in southern Mexico. They believe this is the place where the meteorite crashed into Earth.

Lion's foreleg, bat's wing, and dolphin's flipper

DNA and Evolution

In addition to fossils, scientists also look at DNA to answer questions about evolution. Remember that DNA holds the "life code" of an organism. Scientists can crack parts of this code. Then they can "read" the code to learn about the evolution of an animal. Scientists have found some surprising things. For example some very different animals, such as pigs and mice, have some of the same genes.

Scientists also found that DNA in humans is 98% the same as DNA in chimpanzees! This means that our life codes are very much alike. It also means that humans and chimpanzees probably evolved from common ancestors.

Scientists also study the structures of living animals for clues about evolution. Believe it or not, the bones in a lion's foreleg, a bat's wing, and a dolphin's flipper are very similar. Scientists say that is because these three animals have common ancestors.

Charles Darwin

Charles Darwin played an important role in developing the **theory** of evolution. In 1831, when he was only 22, he sailed around the world. His job was to make a record of the plants and animals he found along the way. His observations on this trip led him to his theory of evolution.

When Darwin returned to England he studied his notes. He also studied the current theories about the origins of various animals, including humans. Then, in 1858, he wrote up his findings. Darwin's theory is the basis of the modern study of evolution.

Charles Darwin

Natural Selection

Remember that when organisms reproduce, the offspring gets half of its genes from each parent. Each parent's genetic contribution controls thousands of traits. So each offspring has a new combination of genes. Over time, some gene combinations lead to survival and reproduction. Other combinations lead to death of the organism. Since dead organisms can't reproduce, those gene combinations are lost. Darwin called this process **natural selection**.

There are four main ideas that explain the process of natural selection.

- **Most organisms have more offspring than can survive**. For instance, an insect lays many, many eggs in her lifetime. But only a few of these eggs will live to become full-grown insects. Dandelions spread thousands of seeds. Only a few of these seeds will land somewhere good for growing into plants.

- **Offspring compete for food and space**. Many or most of an organism's offspring do not survive and reproduce. There is only so much food. There is only so much living space. Only those young organisms that can get food and living space will survive. A seed that lands in a dry, dark place won't grow into a plant. A kitten born in a field where there are five full-grown cats may not get enough mice to live. There is too much competition.

What if both brown and white rabbits lived in a snowy area? Which kind of rabbit would be better protected?

Evolutionary change in one generation may be very small. But over thousands or millions of generations whole new species are formed.

- **Organisms that survive have traits best suited to the environment.** Remember that there is not enough food and space for all organisms. So, which organisms will survive? An organism must be able to get food. It must be able to protect itself from enemies. The strong, fast cats will probably get the mice before the weak, slow ones. Brown rabbits living in the forest are better protected from enemies than white rabbits. The white rabbits are easier to see.

- **Natural selection passes along helpful traits to offspring.** Strong, fast cats will live longer than weak, slow ones. So the strong, fast cats are more likely to have offspring. They will pass these strong, fast traits along to their kittens. Their kittens will have the same useful traits.

Science Practice
Below are some sentences about plants, animals, and their environments. How do you think each organism listed may have adapted to its environment? Write at least one sentence for each on a separate sheet of paper.
- Polar bears live in arctic regions.
- Cacti live in deserts.
- Rats live in cities.
- Ferns live in rain forests.

How Mutation Causes Evolutionary Changes
Scientists today know a lot more about evolution than Darwin did. By studying DNA, they have learned some of the ways that changes in species come about.

Mutations, for example, can cause a species to change. Mutation is different from natural selection. Change by natural selection occurs through new combinations of existing genes. Mutation occurs when a whole new gene is formed. A mutation is a kind of genetic mistake.

Most mutations are harmful. Sometimes, however, a mutation introduces a helpful trait to a species.

Remember that the early horses had four toes. This made them slower runners. Perhaps a mutation caused a horse to be born with fewer toes. This horse, and its offspring, could run from enemies faster. They survived better than those with four toes. Over time, the horses with four toes died out.

People in Science: Barbara McClintock

Barbara McClintock was born in 1902 and spent her life studying corn to learn more about genes. In the mid-1940s, McClintock discovered something shocking. Genes were not fixed in one place—they moved around on chromosomes!

Barbara McClintock

No one believed her. Other scientists laughed at her "jumping genes" idea. For five years, she delivered talks and papers on her "jumping genes," but no one paid any attention. Forty years later, when McClintock was 79 years old, other scientists finally caught up with her. They discovered that genes did jump on chromosomes! This important discovery helped to explain how new species arise and why some cells, like cancer cells, go crazy.

Suddenly, lots of honors came her way. At the age of 81, she won the Nobel Prize.

Chapter Review

Chapter Summary

- Change in species, usually over thousands or millions of years, is called evolution. Scientists study fossils for answers about evolution. Scientists also study DNA and the bone structure of living organisms for clues to the story of evolution.

- Charles Darwin came up with a theory to explain evolution. He said that most organisms have more offspring than can survive. Since offspring must compete for food and space, those with traits best suited to the environment will survive. The survivors will pass their helpful traits along to offspring. This process is called natural selection.

- The study of DNA has added a lot to the study of evolution. Mutations help explain how natural selection happens. A helpful mutation will be passed to offspring. In this way, evolution may happen much faster than Darwin believed.

Chapter Quiz

Write answers to the following questions on a separate sheet of paper.

1. Name two extinct animals.
2. Name two different kinds of animals which have similar bone structures.
3. What are fossils?
4. When did Charles Darwin sail around the world?
5. Give two examples, one plant and one animal, of organisms that have more offspring than can survive.
6. Why won't a seed live if it lands in a dry, dark place?
7. According to Darwin, which organisms will survive?
8. What theory does natural selection explain?
9. How has the study of DNA added to Darwin's theory?
10. Imagine that a mutation causes a fish to be born without fins. Would this help or hurt the fish? Give your reasons.

In Your Own Words

Below is a definition of Darwin's theory of natural selection. Try to explain the theory in your own words. Your teacher will tell you whether to write your explanation or tell it to a classmate.

Natural selection is the way in which organisms best suited to their environments survive and pass their helpful traits along to offspring.

Mad Scientist Challenge

Go to the school or city library. Ask the librarian to help you find a picture of a saber-toothed tiger. This is an animal that is now extinct. Study the picture. Then write several sentences that describe the saber-toothed tiger.

Chapter 10

The Human Body: Cells to Systems

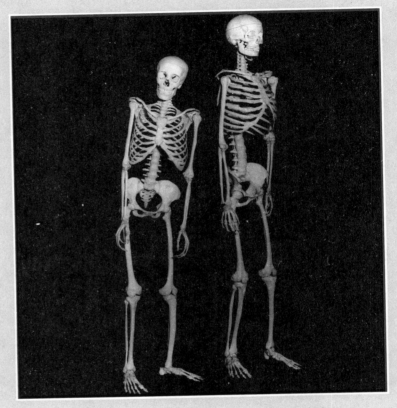

The human body is one of nature's greatest accomplishments. We have thumbs to grasp things. We have big brains to figure things out. Can you think of anything else that makes the human body special?

Chapter Learning Objectives
- Identify two kinds of tissue.
- Name the five senses.
- Name two body organs.
- Describe three important body systems and their functions.

Words to Know

fetus a young mammal after fertilization of the egg but before birth

hormones substances produced by certain glands in the body. These substances are circulated in the blood.

menopause the time in a woman's life when menstruation stops

menstruation the monthly shedding of blood from the uterus when an egg has not been fertilized

neurons cells throughout the body that carry signals to and from the brain (also called *nerve cells*)

organ body tissues that form a working unit, such as a heart or kidney

ovaries the female organs that make egg cells and hormones

puberty the time in life when the reproductive organs develop

system a group of organs working together

tendons tough bands of tissue that attach muscles to bones

testes the male organs that make sperm cells and hormones

tissue a group of cells that all do the same job

uterus the female organ in which a fertilized egg develops into a baby

The human body has a lot in common with other animals. But it also has very important differences. The most important difference is the human brain. It is much larger and more complicated than any other animal's brain. Scientists say the human brain may be the most complicated thing in the universe.

Yet, humans can't run as fast as cheetahs. We can't smell as well as dogs. We can't hear as well as elephants. We can't fly like birds. Many animals have more muscle power than we do.

People make up for all of this with our big brains. We build cars and airplanes to move fast. We make glasses, telescopes, and microscopes to see things that other animals cannot see. We build machines to do some of our muscle work for us. We are even able to build computers that store billions of pieces of information. We send rockets into space and bring them back. And we can write stories and beautiful music.

In this chapter you will read about the human animal. You will learn about many parts of your body and what they do

The Organization of Cells

You have read that animals have specialized cells. Different kinds of cells have different jobs. Cells that do one kind of job are called **tissue**. Blood is a tissue made of blood cells. Muscle is a tissue made of muscle cells.

Groups of tissues that work together are called **organs**. The heart is a good example of an organ.

Of course having big brains gets us in trouble sometimes, too. Can you think of problems the human species might not have if we weren't quite so smart?

Blood tissue and muscle tissue work together to help the heart do its job. The stomach, lungs, eyes, and ears are also organs.

Organs are grouped together in **systems**. The *circulatory system* moves blood. The *digestive system* breaks down food into simpler parts so that body cells can use it. The *respiratory system* takes in oxygen and combines it with food to get energy. The *nervous system* carries messages all over the body.

The Five Senses

Humans have five senses: sight, hearing, smell, taste, and touch. Each of these senses has an organ. All of these sense organs belong to the nervous system.

Eyes are the organs of sight. Light bounces off objects and enters the eye. Then the light triggers nerves in the back of your eye. When these nerves send messages to your brain, you see an image.

Many people depend more on their sense of sight than on their other senses. However, blind people learn to depend on their hearing and touch. These other senses become much sharper when the sense of sight is taken away.

Ears are the organs of hearing. All sounds are made by vibrations. Think of a musical instrument. When you pluck a string, you make it vibrate. This makes sound. The vibrations can move a long way. Moving sound vibrations are called *sound waves*. These waves are really molecules of air bumping into one another.

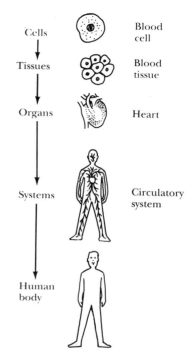

All the systems put together make the human body.

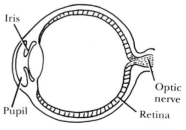

The human eye

Rabbits have great hearing. That's partly because they have such big ears. Why do you think people sometimes cup their hands behind their ears to hear better?

Your sense of smell is much stronger than your sense of taste. Often when you think you are tasting something you are really smelling it.

The ears stick out and catch sound waves. Tiny bones inside the ear are moved by the vibrations. Nerves pick up the movement of the tiny bones and send a patterned signal to the brain. Then you hear a sound.

The tongue is the organ of taste. People can only taste four flavors: sour, sweet, salty, and bitter. Different areas of the tongue taste different flavors. The tip of the tongue tastes sweet and salty things. The sides of the tongue taste sour things. And the back of the tongue tastes bitter things. Not all animals have the same sense of taste. Cats, for example, don't taste sweet things at all.

The nose is the organ of smell. Suppose you are smelling an onion. Tiny molecules come off the onion you are smelling. These molecules drift up to your nose. There, the molecules from the onion trigger nerve cells.

Skin is the organ of touch. Certain areas of skin have lots of nerve cells and are very sensitive to touch. Other areas are less sensitive. Fingertips are one sensitive area. The sense of touch protects the body from harm. If something sharp or hot touches your body, you know to move away. The sense of touch works like an alarm system.

Skin is one of the human body's most important organs. It protects your body by keeping it from drying out. It also keeps bacteria and other harmful organisms away from the rest of your body. And your skin helps keep your body temperature at the right level.

Skin has sweat glands in it. A gland is an organ that makes a certain substance. Sweat glands make sweat. The main purpose of sweat is to help keep the body at a comfortable temperature. The wetness of sweat cools the skin when you get overheated.

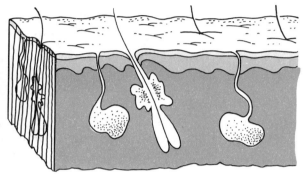

The skin protects the body and keeps its temperature constant.

Science Practice

Write answers to the following questions on a separate sheet of paper.

1. What do you think eyebrows protect? How do they do this?
2. What do you think is the purpose of the hair on your head?
3. What would you be unable to do if you didn't have fingernails?
4. Do you think toenails have any purpose? If yes, what? If no, why do you think people have them?
5. Below are four kinds of tissue. Which of these do you think are found in the eye? In the stomach?

   nerve tissue   bone tissue
   blood tissue   muscle tissue

The Nervous System

Have you ever wondered why humans can read when no other animals can? Well, no other animal has a brain like the human brain. The human brain is larger and more developed than any other animal's.

The brain is the control center of the human body. It is made up of billions of nerve cells, or **neurons**. The brain sends messages to all parts of the body. It also receives messages from all parts of the body. The brain controls many body functions, and stores information, too.

Messages go downward from your brain to your *spinal cord*. The spinal cord is like a big rope of neurons. It runs down the inside of your back bone. The spinal cord connects to neurons all over your body. Messages travel from one neuron to the next until they reach their destinations.

The brain, spinal cord, sense organs, and neurons all work together to make up the *nervous system*.

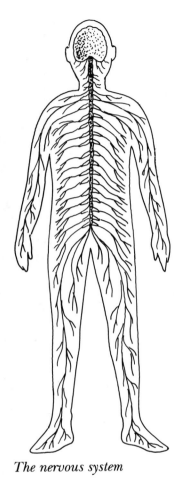

The nervous system

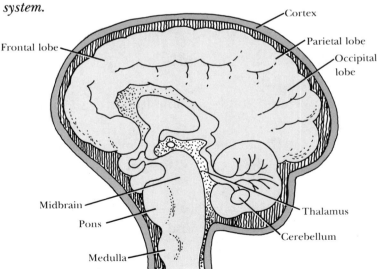

The brain is very complicated. Scientists are just beginning to sort out its many mysteries.

The Support System

A human body has 206 bones in it. These bones work together as your *skeleton*. The skeleton gives your body support. You would not be able to sit, walk, or stand without it.

The skeleton also protects many of your most important organs. It makes a cage around your heart and lungs. Another part of your skeleton forms a hard shell that protects your brain.

Bones are made of living cells and also a mineral called *calcium*. The places where bones meet are called *joints*. You have joints at the wrist, knee, and on your fingers. There are many other joints on your body. Can you think of three more joints?

Muscles Help You Move

Flex a muscle. You probably just flexed an arm muscle. That muscle is attached to your arm bones by **tendons**. Tendons are tough bands of tissue that grip both muscle and bone. Muscles help your skeleton to move the way you want it to move. These are called *voluntary muscles*.

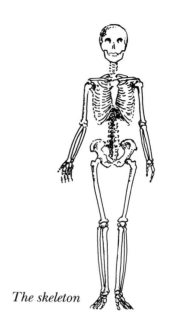

The skeleton

There is a lot of calcium in milk. People who are still growing should drink plenty of milk. It helps their bones grow and stay strong.

Healthy, older women should take in 800 milligrams of calcium per day to avoid osteoporosis, a disease in which bone matter is lost. But most women take in far less than that. 24 million Americans, mostly women over 50, suffer from osteoporosis.

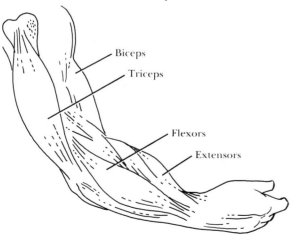

The arm muscles are attached to the bone by tendons.

Chapter Ten

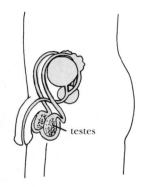

Male reproductive system

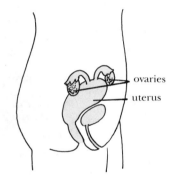

Female reproductive system

Some muscles work without bones. When you swallow food, certain muscles push the food down to your stomach. Other muscles cause blood to move throughout your body. These muscles are not attached to bones. They are called *involuntary muscles.*

Reproduction and Growth

The human reproductive system does not fully develop until **puberty**. During puberty, certain glands begin producing **hormones**. These are substances that are circulated in the blood and that cause changes in the body.

For instance, during puberty girls develop breasts and broader hips. Boys develop broader shoulders and more muscles. Both boys and girls grow hair under their arms and around their reproductive organs. Boys' voices get lower. Puberty for girls happens between the ages of 10 and 13. It happens for boys between the ages of 13 and 16.

During puberty both males and females begin producing sex cells. Male sex cells, called *sperm*, are made in the **testes**. An adult male makes millions of sperm cells a day. Female sex cells are called *eggs*. Egg cells are made in the **ovaries**. Adult females make only one egg cell a month.

Each time the ovaries of a woman make an egg cell, the uterus prepares for a fetus. The **uterus** is the muscular organ in which a fetus develops. A **fetus** is an unborn baby. Each month the walls of the uterus become thick with blood. This blood will help nourish a developing fetus. Because the uterus is made of muscle, it can stretch a lot. If an egg cell has been fertilized by a sperm cell, the fetus begins growing in the uterus.

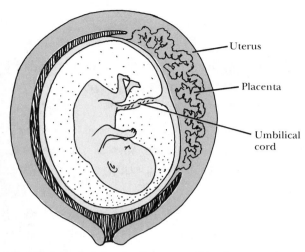

Most fetuses develop in the uterus for nine months before emerging as babies.

If the egg is not fertilized, the extra blood lining the uterus leaves the body. This is called **menstruation**.

When a woman grows older, menstruation stops. This usually happens at about the age of 50. The stopping of menstruation is called **menopause**.

People in Science: Dr. Bernadine Healy

As the first woman to head the National Institute of Health, Bernadine Healy is making sweeping changes. She believes that health research has ignored diseases that affect mostly women, such as osteoporosis. She wants to close the "knowledge gap" between women's and men's health. Not everyone likes the changes that Healy is making. But she says, "I am willing to go out on a limb, shake the tree, and take a few bruises."

Dr. Bernadine Healy

Chapter Review

Chapter Summary

- Cells that work together to do the same job are called tissues. Different kinds of tissues work together to form organs. And organs work together to make systems.

- The five senses are sight, hearing, smell, taste, and touch. The eyes control sight. The ears control hearing. The nose controls smell. The tongue controls taste. The skin controls touch.

- The nervous system is made up of the brain, spinal cord, sense organs, and neurons. The job of this system is to carry messages. The brain controls the nervous system.

- The skeletal system supports your body. It also protects important organs, such as the lungs and heart. The muscle system helps you move.

- At puberty, the reproductive system develops. Hormones cause males to begin producing sperm. Females begin producing egg cells once a month.

- The uterus is the female organ in which a fetus develops. Each month the walls of the uterus fill with blood to prepare for a fetus. If an egg cell is not fertilized by a sperm cell, a fetus won't begin developing. The blood from the uterus leaves the body. This is called menstruation.

- The time when a woman stops menstruating, usually at about 50, is called menopause.

Chapter Quiz

Write answers to the following questions on a separate sheet of paper.

1. Name two kinds of bodily tissues.
2. Name two organs. What do they do?
3. Name two body systems. What do they do?
4. Name the five senses and the organs that go with each sense.
5. Which body organ helps keep you at a steady temperature? How does it do this?
6. Where are most of your nerve cells?
7. Name two important functions of your skeleton.
8. Name one kind of voluntary muscle. Then name one kind of involuntary muscle.
9. What chemical substances cause changes in the body at puberty?
10. What are male sex cells called and where are they made? What are female sex cells called, and where are they made?

Mad Scientist Challenge: Drawing the Body Systems

On a separate sheet of paper, make two copies of the human figure on this page. Draw the skeletal system on one. Draw the nervous system on the other. Use the list below to label each system.

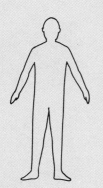

skeletal system	nervous system
skull	brain
ribs	spinal cord
arm bones	neurons leading to arms and legs
leg bones	

Chapter Ten 117

Chapter 11

The Human Energy Systems

The respiratory system, circulatory system, and digestive system all work together.

Chapter Learning Objectives
- Describe the path and organs of the circulatory system.
- Describe the path and organs of the digestive system.
- Describe the path and organs of the respiratory system.

Words to Know

arteries blood vessels that carry blood away from the heart

capillaries tiny blood vessels that connect arteries and veins

circulatory system the system that carries blood through the body, delivering food and oxygen to all parts of the body

digestive system the system that breaks down food so that body cells can use it

enzymes substances that cause chemicals to change form in the body

esophagus the tube that carries food from the mouth to the stomach

feces wastes expelled from the body

plasma the liquid part of blood

platelets solids in the blood that help stop bleeding at an injury

red blood cells blood cells that carry oxygen and carbon dioxide

respiratory system the system that takes in oxygen and combines it with food to produce energy

veins blood vessels that return blood to the heart

white blood cells blood cells that fight bacteria

Think about what you had for breakfast this morning. If it is still morning, most of that food is probably still in your stomach. If it is afternoon, your breakfast is probably already in your bloodstream giving energy to your cells.

In this chapter you will learn how your body converts food and oxygen into energy. This is done by the digestive and respiratory systems. You will also learn how the circulatory system distributes that energy throughout the body.

The Circulatory System

To circulate means to move around. The **circulatory system** moves blood around the body. The blood delivers food and oxygen to all the body's cells. It also carries wastes away from the cells.

The most important organ of the circulatory system is the heart. The heart is a muscle about the size of your fist. It squeezes together to pump blood to every part of the body. The blood leaves the heart through blood vessels.

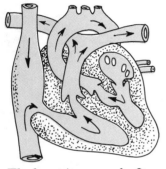

The heart is a muscle. It pumps blood to every part of your body.

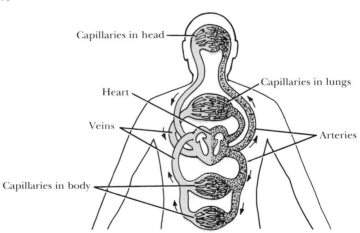

The circulatory system moves blood around the body.

Science Alert

Heart disease kills hundreds of thousands of Americans each year. A heart attack happens when blood vessels become clogged. If a blood vessel is clogged, blood cannot get through. Soon the body does not get the oxygen from the blood it needs to stay alive.

Doctors are learning more every day about how to prevent heart disease. Not smoking is the best way to protect yourself from heart disease. Cutting back on fatty foods is another good way. Certain kinds of fat can build up on the walls of the blood vessels. This makes less room for the blood to get through. Sooner or later, the vessels may get clogged.

Getting plenty of exercise also helps keep your blood vessels open. When you exercise, your heart works hard. It pumps lots of blood through the circulatory system. This keeps the blood vessels open wide.

Bruises are caused by broken blood vessels. If you fall and hit your arm on a rock, many tiny capillaries will break. Blood spills out of them. The blood is trapped under your skin and causes the skin to darken.

Compare these two blood vessels.

There are three kinds of blood vessels. **Arteries** carry blood *away* from the heart to the body cells. **Veins** *return* blood to the heart. **Capillaries** connect arteries and veins. Capillaries are the smallest blood vessels in the system. Some capillaries are so small that only one blood cell can pass through at a time.

Blood is made up of both liquids and solids. **Plasma** is the liquid part of blood. Red blood cells, white blood cells, and platelets are solids in blood. **Red blood cells** carry oxygen and carbon dioxide throughout the circulatory system. **White blood cells** fight off

Chapter Eleven 121

bacteria and sickness in your body. When people get sick, the body produces many more white blood cells than usual. **Platelets** help stop bleeding in injuries. They group together to close off a cut.

You may have heard of *high blood pressure.* Your blood pressure is the measure of how hard your blood has to push to get through the vessels. If your blood vessels are open wide, the pressure is lower. If your vessels are closed in, the pressure is greater. High blood pressure can lead to heart disease.

The Digestive System

You know that blood delivers food to all the cells in your body. But how can blood carry a fried egg? First the egg must be broken down into tiny pieces. In fact, it must be broken all the way down into molecules. This is the job of the **digestive system**.

First you take a bite of the egg and chew it. This helps break it down into smaller pieces. Next, an enzyme in your saliva helps break these pieces down further. **Enzymes** are substances that bring about chemical changes. In digestion, enzymes help break up food.

People often say, "Chew your food well before you swallow." Do you think this is good advice? Explain why or why not.

After you chew the food, you swallow it. It goes down the **esophagus**. This is a long tube leading from your throat to your stomach. Muscles along the esophagus help push the food down.

Your stomach has muscles in it, too. These muscles move in a wave-like motion. This movement stirs up the food and breaks it down even more. There are more enzymes in the stomach, too. These also help break the food down.

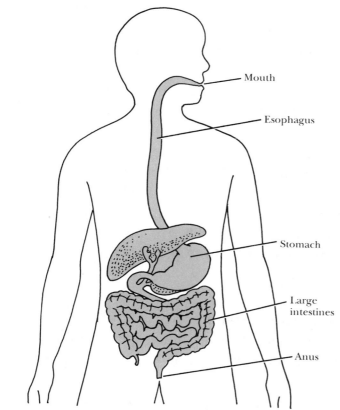

The digestive system turns food into fuel your body can use.

From the stomach, food moves into the small intestine. By now your egg is just a lot of molecules ready to feed your cells. These molecules pass through the wall of the small intestine and into the bloodstream. The bloodstream carries them all over the body.

Not all of the food you eat gets turned into energy, however. Some of it just isn't useful. These wastes are called **feces**. They pass from the small intestine to the large intestine. Then they pass out of the body through the anus.

Chapter Eleven   123

Science Practice

Choose one answer for each question. Write your answers on a separate sheet of paper.
1. Arteries are blood vessels that (a) carry blood away from the heart (b) return blood to the heart (c) connect capillaries and veins.
2. White blood cells (a) help stop bleeding (b) are the liquid part of blood (c) fight off bacteria and sickness in the body.
3. Enzymes are substances that help (a) move food to the stomach (b) make chemical changes happen (c) stop bleeding.

The Respiratory System

Remember that all living things carry out respiration. That is the way that cells use food and oxygen to get energy. You just read about the digestive system and how it gets food to all your cells. Now you will read about how the **respiratory system** gets oxygen to all your cells.

When you take a breath you draw air into your body. Usually you breathe in through your nose. But you can also breathe in through your mouth. From there the air goes down your throat to a box-like structure called the *larynx*, or voice box. Then the air goes down the *trachea*, or windpipe. Branching from the trachea are two tubes called *bronchi*. Each bronchus enters one of the lungs.

Remember that air is made of oxygen, nitrogen, and other gases.

Do you remember where you get the oxygen you breathe? The oxygen in the air comes from plants. Plants put out the oxygen as a waste product of photosynthesis. This is a good example of how very different forms of life can depend on each other.

Your lungs are something like sponges. They have many little air sacs in them. The air sacs are covered with tiny capillaries. When you inhale, the air sacs fill with air. The capillaries are exposed to the air. Oxygen in the air passes through the capillaries and into your blood.

The blood then carries the oxygen to all the cells in your body. The oxygen combines with the food in the cells and energy is released. The waste products of respiration are carbon dioxide and water. These wastes are carried through the circulatory system to the lungs. When you breathe out, you expel carbon dioxide and water vapor.

Think about how the human body might help plants. Remember that plants need four things for photosynthesis. You breathe one of those things into the air with every breath. Which one?

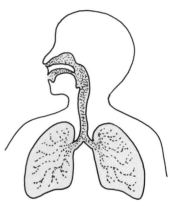

The respiratory system gets oxygen to your cells.

People in Science: Antonia Coello Novello

Antonia Coello Novello is the first woman and the first Hispanic to serve as United States Surgeon General. She reported that more than 3,000 teenagers become regular smokers each day. In the United States 390,000 people die every year from diseases of the heart and lungs caused by smoking. Cigarette smoke damages the lungs. Over time it can prevent your body from getting the oxygen it needs.

"Because only a very small percentage of smokers begin smoking as adults," said Novello, "efforts at prevention must focus on children."

Antonia Coello Novello

Chapter Review

Chapter Summary

- The circulatory system moves blood around the body. Blood carries food and oxygen to all the body cells. It also carries wastes away from the cells. The heart is the main organ of the circulatory system. Blood vessels that carry blood away from the heart are called arteries. Blood vessels that return blood to the heart are called veins. Capillaries are tiny blood vessels that connect arteries and veins.

- The digestive system breaks down food so that it can get into the bloodstream. The mouth, esophagus, stomach, small intestine, and large intestine help in this process.

- The respiratory system brings oxygen into the bloodstream. The nose, larynx, trachea, bronchi, and lungs are parts of this system. Once oxygen reaches the lungs, it can be absorbed into capillaries. Carbon dioxide and water vapor are breathed out of the lungs.

Chapter Quiz

Write answers to the following questions on a separate sheet of paper.

1. Describe the three different kinds of blood vessels.
2. Name three solids that are in blood. Describe their jobs.
3. What are three steps you can take to avoid getting heart disease?
4. Is high blood pressure good or bad? Explain your answer.
5. Where does food go after you have chewed it?
6. Into what organ do undigested foods go?
7. Why do your cells need oxygen?
8. What system gets oxygen to the bloodstream?
9. Where do the bronchi lead?
10. What two wastes do you breathe out into the air?

Mad Scientist Challenge: Drawing the Body Systems

On a separate sheet of paper, make three copies of the human figure on this page. Draw the circulatory system on one. Draw the digestive system on one. And draw the respiratory system on the last. Include and label your drawings with the parts listed below.

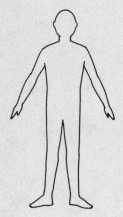

circulatory system	digestive system	respiratory system
heart	mouth	nose
veins	esophagus	trachea
arteries	stomach	bronchi
capillaries	small intestine	lungs

Chapter Eleven 127

Chapter 12

Healthy Living

This athlete takes good care of her body. She gets plenty of exercise. She makes sure to eat nutritious food, too. Do you think she smokes cigarettes?

Chapter Learning Objectives
- Explain what a disease is.
- Name four effective ways to help your body stay healthy.
- Explain how cigarettes harm the body.

Words to Know

carbohydrates sugars in food that give people energy

disease sickness; a condition in which the body is not functioning properly

fats nutrients in foods that supply the body with fuel or energy

minerals substances found in non-living things; people need minerals in their diets in order to stay healthy

nicotine a substance in cigarettes that can cause heart disease and lung cancer

nutrients substances in foods that body cells need

proteins nutrients in foods that build body tissues

virus a microscopic "organism" that causes diseases and is missing some cell parts; *viruses* can grow and reproduce only in certain living cells

vitamins substances found in many foods; people need vitamins in their diets in order to stay healthy

---

In the 1300's a sickness swept through Europe. Almost one quarter of the population died of the disease in only 20 years. It was called the plague, or "Black Death." No one could figure out exactly where the plague came from or how it got passed along.

Now we know that the plague was caused by bacteria. Since the 1300's scientists have learned a

The plague was caused by bacteria that infected rats. Fleas that bit the rats then spread the bacteria to humans. The disease is still around in other parts of the world. But medicines now available keep many people from dying from it.

Chapter Twelve   129

lot about what makes people sick. They have also learned a lot about what keeps us healthy. In this chapter, you will learn about some of the things that cause disease. You will also learn some good ways to help yourself stay healthy.

What Is Disease?

A **disease** is a kind of illness or sickness. Even the common cold is a disease, though not a very serious one. Most diseases are caused by tiny organisms, such as bacteria. People catch diseases from other animals. Mosquitoes, fleas, and pigs can carry disease. Humans also pass diseases to one another.

The human body has many ways to fight disease. The skin is one way. It keeps harmful microscopic organisms away from most of our organs. The hair in our noses also keeps harmful organisms out of our bodies. The hair filters the air we breathe.

But some bacteria still find their way into the body. The white blood cells then go to work to kill the disease. Suppose for example that bacteria cells enter your blood stream. White blood cells will surround the bacteria and try to kill it. But sometimes the bacteria reproduce faster than the white blood cells can handle. In that case, your body would begin to produce more white blood cells. In the meantime, however, you would get sick.

The defenses against disease work best when a person is already healthy. If you eat well and get enough exercise and sleep, there's less chance of your getting sick.

Why do you think it is important to cover your mouth when you cough or sneeze? Why do you wash your hands after using the bathroom?

Viruses Cause Many Diseases

A **virus** is smaller than a cell. It has a center made of genetic material and an outer coat of protein. Viruses cause many diseases. AIDS, flu, colds, polio, chicken pox, measles, and mumps are all caused by viruses.

Viruses are very mysterious. Scientists still don't completely understand them. In fact, they are not even sure whether viruses should be called living or non-living!

By themselves, viruses appear to be non-living. They do not carry out the normal life functions of cells. But viruses come to life by invading living cells. Once in a cell, viruses use parts of the cell to reproduce and get energy. Once they have used a cell this way, they usually destroy the cell.

Most diseases caused by viruses can now be prevented by doctors. But some are still deadly.

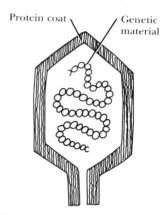

Viruses are composed of genetic material surrounded by protein.

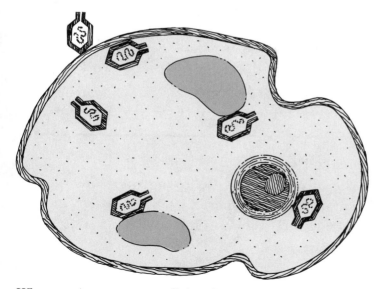

When a virus enters a cell it takes over parts of the cell to reproduce and get energy.

Do you think it would be correct to call a virus a parasite? Why, or why not? Look back at page 65 if you need to refresh your memory about parasites.

On the Cutting Edge

AIDS is a disease that harms the body's ability to fight other diseases. Since the disease appeared in 1981, over 10 million people have been infected. Many people have died of AIDS. There is no known cure.

However, there are known ways to protect yourself from AIDS. AIDS is passed through bodily fluids. You *cannot* get it from mosquitoes, shaking hands, or being near to or touching someone with AIDS. AIDS is passed along in four ways: 1) through sexual contact; 2) by sharing needles used for drugs; 3) through blood given to a person by another who was infected; 4) from a pregnant woman who has AIDS to her unborn child.

Scientists recently have developed a vaccine for AIDS that works on monkeys. They do not yet know, however, whether the vaccine would work the same way in people. Scientists are getting closer, but they have a lot more work to do in finding a cure for AIDS.

Science Practice

Write answers to the following questions on a separate sheet of paper.
1. Name three defenses the body has against disease.
2. What must viruses do before they can reproduce?
3. What are the four ways you can get AIDS?

Nutrition Means Eating Well

Have you ever heard the expression "You *are* what you eat"? There is certainly some truth to that. If you eat well, you are likely to feel well. Eating well is also one of the best ways to avoid disease.

Nutrients are the substances in foods that everyone needs to stay healthy. There are six main kinds of nutrients. These are carbohydrates, fats, proteins, water, vitamins, and minerals. If any one of these nutrients is missing from your diet, your health may be in danger.

Carbohydrates are sugars and starches. They give you energy. Fruits, vegetables, grains, and potatoes are all good sources of carbohydrates.

Fats are also a source of energy. But your body does not use them as efficiently as it uses carbohydrates. Most people eat more fat than they should. Fat can build up on blood vessel walls and lead to heart disease. Butter, mayonnaise, and meat are all sources of fat.

Water is also an important nutrient. All of the cells of the body contain water.

Proteins are tissue builders. Almost every part of your body is made of protein. This includes your hair, fingernails, blood, muscles, and organs. Meat, fish, nuts, beans, and dairy products are all good sources of protein.

Vitamins are substances found in tiny amounts in plants and animals. To stay healthy, people need certain vitamins in their diet. **Minerals** are non-living substances, such as iron and calcium. People also need small amounts of minerals in their diet.

By eating from each of the five basic food groups every day, you will get the nutrients you need. The five food groups are different from the six main nutrients. Look at Appendix E for examples of each food group.

Many people take vitamin tablets. Some scientists believe these supplements can make you healthier. But others say that a well-rounded diet gives you all the vitamins you need.

Look at Appendix F and Appendix G for more information about minerals and vitamins.

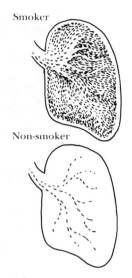

Compare the lung of a smoker with the lung of a non-smoker.

Science Alert

You have probably heard people talking about cholesterol. This is a substance found in some foods and in the human body. You need small amounts, but too much cholesterol may be unhealthy. Cholesterol causes fat to build up in the arteries. It is a good idea to limit the amount of cholesterol in your diet. To do this, cut back on red meat, eggs, and high-fat foods.

Did you know that most Americans eat too much salt? Doctors believe that salt is a major cause of high blood pressure. To prevent trouble, it is a good idea to eat as little salt as possible.

Other Ways to Stay Healthy

When it comes down to it, no one can tell you what to eat. Or what not to eat. But it helps to be informed so you can make good decisions for yourself.

Heart disease is the number one killer disease in the United States. Good nutrition is important for protecting yourself against heart disease. Getting exercise and not smoking will also help protect you from this kind of illness.

Nicotine is a substance in cigarettes that causes blood vessels to narrow. Narrowed blood vessels make it difficult for blood to flow freely through the body. Over time this narrowing can lead to heart disease. Smoking also causes many respiratory diseases. For instance, lung cancer can be caused by smoking cigarettes. Cancer is the number two killer disease in the United States.

Exercise keeps your muscles in good working order. Your heart is a big muscle. It needs exercise just as other muscles in your body do. By exercising, you keep your heart strong. Exercise also keeps your blood vessels open wide. This helps prevent heart disease.

On the Cutting Edge

Do you like your body? Most Americans, especially girls and women, answer "No" to that question. Many feel that they are too fat. So many, in fact, that dieting is an 8.4 billion dollar business in the United States!

Scientists are discovering, however, that dieting is bad for you. At least 90 percent of people who lose weight on diets gain it back again. What happens is that your body believes it is starving. So it burns fewer calories. When you go off the diet, the body protects itself by storing even more calories as fat. In other words, the more you diet, the fatter you are likely to get!

Some scientists believe that we have genes to make us certain sizes. Just as people are short and tall, they are fat and thin. If you really want to lose weight, and keep it off, do it *very* slowly. You don't even have to go on a "diet." Instead, just eat healthy foods—fruits, vegetables, and carbohydrates. Cut out as many fatty foods as you can. The best way is to get lots of exercise. Exercise increases the calories your body burns, even at rest.

Chapter Review

Chapter Summary

- A disease is anything that harms the body's health. Many diseases are caused by small organisms, such as bacteria. Others are caused by viruses.

- The skin is a good defense against disease. It helps to keep harmful organisms and viruses out of our bodies. Once a disease enters the body, the white blood cells work at killing it. Even so, all people get sick sometimes. This is more likely to happen when people are not eating or sleeping well.

- Viruses cause many diseases. Scientists are not sure if viruses should be called living or non-living. Viruses invade living cells. The viruses then use the parts of the invaded cell to carry out their own life functions. Viruses reproduce very quickly.

- Eating well is one of the best ways to avoid disease. You need six kinds of nutrients in your diet to stay healthy: carbohydrates, fats, proteins, water, vitamins, and minerals.

- Avoiding smoking cigarettes is another easy way to stay healthy. Getting plenty of exercise will help keep your heart and other muscles strong.

Chapter Quiz

Write answers to the following questions on a separate sheet of paper.
1. Name three diseases and their causes.
2. How does skin help the body fight disease?
3. Name five diseases caused by viruses.
4. Which virus harms the body's ability to fight other diseases?
5. What are the six kinds of nutrients you need for a healthy diet?
6. What do carbohydrates do for your body?
7. What do proteins do for your body?
8. Which disease kills the most people in the United States? Which disease comes in second?
9. Name two of the diseases that smoking cigarettes can lead to?
10. How does exercise help you to stay healthy?

Charting a Healthy Life

Make a health chart on a separate sheet of paper. Draw a line down the center of the paper. Write "Don't" on the top left part of the paper. To the right of the line write "Do." Under the "Don't" heading, write at least four things that are bad for your health. Under the "Do" heading, write at least four things that will help you keep good health.

Chapter 13

Living Things Depend on Each Other

The organisms in this picture depend on each other. The farmer feeds the cows. The cows "feed" the farmer. And they all depend on the environment.

Chapter Learning Objectives
- Explain what an ecosystem is made up of.
- Describe the food, water, and oxygen and carbon-dioxide cycles.
- List three reasons for conserving natural resources.

Words to Know

community all the organisms that live in one habitat

conservation the wise and careful use of our natural resources

consumers organisms that eat other organisms

decomposers organisms that eat dead matter

ecosystem a system that is formed by the interaction of a community of organisms and its environment

evaporation the process by which heat changes water to water vapor

food chain a group of organisms, each of which is dependent on another for food

fossil fuels fuel products made from plant and animal remains, such as coal, petroleum, and natural gas

habitat the place where an organism lives

natural resources substances found in nature that are useful to humans

population the group of one species living in a certain place

producers organisms that make their own food

recycle to use something over and over again

solar energy energy produced by the sun

Take a breath of air. It is possible that some of that air was breathed by Columbus! Or maybe even Cleopatra! The air our bodies use has been around for millions of years. We breathe it over and over again. In fact, all the air, water, and food your body takes in has been used before. Reusing substances is called **recycling**.

In this chapter you will see how nature recycles its parts. You will also learn how different living things depend on each other.

Groups of Organisms

A **habitat** is a place where an organism lives. The word habitat can be used to describe a small place or a big place. For example, a bird's habitat is the tree in which it builds a nest. But the entire forest where it lives can also be called its habitat.

A group of organisms of one species, all living in a certain habitat, make up a **population**. The frogs in one pond are a population. The blackberry bushes in one forest are a population. The people in a city are a population.

Every living environment is home to many populations. A group of different populations living in one place is called a **community**. All the plants and animals living in the desert make up the desert community. All the organisms living in a mud puddle make up the mud puddle community.

All the organisms in a community interact with one another. Animals eat plants and other animals. Plants provide animals with oxygen. Dead organisms provide bacteria with food.

All living things depend on one another and their environments. The study of the relationship between different populations in a habitat is called *ecology*. The community and the environment combined form an **ecosystem**. Like the systems in the human body, ecosystems are made of many parts working together.

Chemical fertilizers, human wastes, and soap all go into our water supply. Do you think this is dangerous to the organisms that depend on water?

When an ecosystem is working smoothly it is said to be balanced. A balanced ecosystem supports and maintains the organisms in it.

Ecological Change

Over time, ecological communities can undergo changes. Think about a pond community, for example. Throughout the years, leaves and dead trees fall into the pond. Soil washes into it, too. As it fills up, the pond becomes more and more shallow.

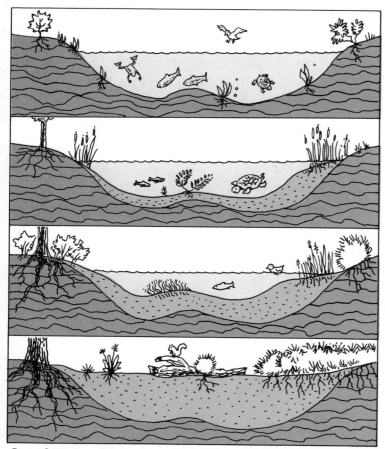

Over the years, this environment changed from a pond into a forest.

Fish die and sink to the bottom. The fish decompose and make the soil on the bottom of the pond rich for plant life. More and more plants grow in the pond. Slowly the pond fills in. After a time, it becomes a meadow.

Mice and grasses move onto the meadow. Then bigger bushes grow. Finally, trees take over. What was once a pond has become a forest.

The Food Cycle

All organisms are either producers or consumers. Green plants, and any other organisms that have chlorophyll in their cells, are **producers**. They make their own food from sunlight, water, and carbon dioxide. This food feeds all other life on Earth.

Animals, fungi, some protists, and bacteria are **consumers**. They eat the food that producers make. Or they eat the animals that eat the food that producers make.

The pictures in the margin show a **food chain**. A food chain is a group of organisms. Each of the organisms in a food chain eats another one of the organisms. Every food chain begins with a producer. In the ocean, algae are at the bottom of the food chain. Little fish eat the algae. Bigger fish eat the little fish.

However, most consumers eat several kinds of food. So the food chains cross over. These more complicated kinds of food chains are called *food webs*.

Third-level consumers

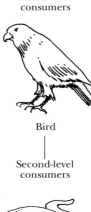

Bird

Second-level consumers

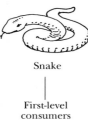

Snake

First-level consumers

Mouse

Producers

Plant

Food chain

Some consumers, such as bacteria and fungi, are called **decomposers**. To decompose means to break down. Decomposers feed on dead organisms. They help to break down the dead matter so it can return to the soil as nutrients. Plants take in the nutrients and the food cycle goes on.

Energy Sources

When you drive a car, you are recycling dead plants and animals. Coal, petroleum, and natural gas are called **fossil fuels**. Fossils are the remains of plants and animals. Fossil fuels are made of organisms that died millions of years ago. Deep in the Earth, chemical changes slowly turned them into useful fuel. People discovered that they could burn these fossils for energy.

Many people believe that it is not wise to use all the fossil fuels in the Earth. They believe other kinds of energy should be found. Energy from the sun, for example, can never be used up. So, scientists are working on ways to trap the sun's energy. Energy from the sun is called **solar energy**. Many people already use solar energy to heat their homes.

Science Practice

On a separate sheet of paper, draw a food chain. Include these three organisms.

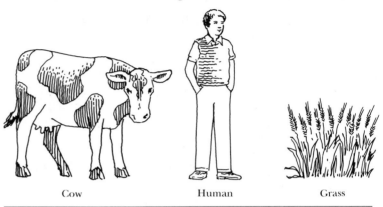

Cow   Human   Grass

The Water Cycle

There is only a certain amount of water on Earth. People use that water over and over again. Rain water falls onto the ground. Then ground water runs into streams, and the streams run into big rivers. The big rivers run into the oceans. Somewhere in the water cycle humans and other organisms use the water. But eventually even the water that is used returns to the water cycle.

Evaporation is the process by which water turns to water vapor when it is heated. The sun causes water to evaporate off the Earth and oceans. It rises into the sky where it collects in clouds. Eventually it falls to Earth again as rain or snow.

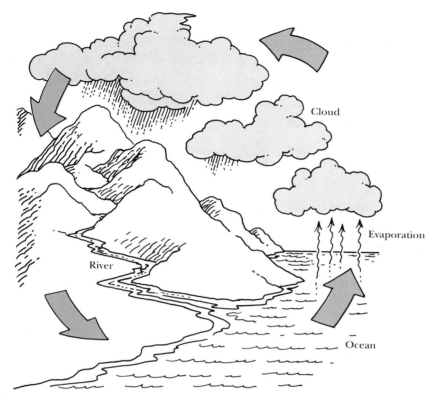

The water cycle. Water falls from the clouds, makes its way to the ocean, and evaporates back up into the sky.

The Oxygen and Carbon-Dioxide Cycle

You already know about this cycle. It has to do with photosynthesis and respiration. When green plants carry out photosynthesis, they give off oxygen as a waste. People and other animals use this oxygen for respiration.

Remember that respiration is the way all plants and animals get energy from oxygen and food. A waste product of respiration is carbon dioxide—a chemical that plants need for photosynthesis. As you can see, the cycle goes on and on. Plants and animals need one another.

The oxygen-carbon dioxide cycle. Oxygen and carbon dioxide are constantly being recycled by plants and animals.

Conservation of Natural Resources

Water, air, soil, minerals, forests, wildlife, and fossil fuels are all **natural resources**. Natural resources are substances found in nature that are useful to humans. Of course things can be useful in many different ways. Oil is useful for energy production. Clean air and water are necessary for our health.

But some natural resources are useful to us simply because they are so beautiful.

Today there are more people on Earth than ever before. Many of our natural resources are getting used up. Forests have been chopped down. Many species of plants and animals have been wiped out. The loss of any species of organism is a very serious matter. It affects all the other populations in that organism's habitat. The end of one species can throw a whole ecosystem out of balance.

Conservation is the wise and careful use of natural resources. More and more people are realizing how delicate the Earth's ecosystem is. These people, called *conservationists*, argue that we must save the Earth's natural resources for the future. They say we must use the air and water in ways that keep it clean. They also say we must respect other species and preserve our planet's ecosystem.

On the Cutting Edge

Ever see big piles of old tires by the roadside? Some 285 million tires are thrown away each year. But what can you do with an old tire? Grind it up, add it to the roadway, and drive on it some more!

A new federal law now requires that 5 percent of roads laid using federal money in 1994 must contain scrap rubber from old tires. That will account for more than 3,000 miles of roadway. The law increases the percentage over time. Many states are thinking about passing similar laws. People must continue to think of creative ways to recycle the things we use.

Chapter Review

Chapter Summary

- A habitat is where an organism lives. A population is all the organisms of one species in a certain place. A community is all the organisms, including all species, in a certain place. Ecology is the study of how organisms depend on one another and their environment.

- Producers are organisms with chlorophyll in their cells. Consumers eat producers. Some consumers also eat other consumers. Decomposers eat dead plants and animals. In doing this, decomposers return nutrients to the soil for producers to use.

- All water on Earth is recycled. It falls as rain. Then it washes into rivers and the ocean. There it evaporates up into the sky. Finally, it falls as rain once again.

- The oxygen and carbon-dioxide cycle shows how plants and animals are dependent upon one another. In photosynthesis, plants give off oxygen as a waste. Animals use this oxygen for respiration. In turn, animals give off carbon dioxide as a waste of respiration. The plants use this carbon dioxide in photosynthesis.

- Water, air, soil, forests, wildlife, minerals, and fossil fuels are all natural resources. It is unwise to use natural resources carelessly.

Chapter Quiz

Answer the following questions on a separate sheet of paper.

1. Some people return their newspapers and bottles so that they can be used again. What is this called?
2. Name five different kinds of organisms that live in your community.
3. Are you a producer or a consumer in the food cycle?
4. How do decomposers help plants?
5. What causes water to evaporate off the oceans?
6. Why is water pollution a serious problem?
7. How do plants help animals in the oxygen and carbon-dioxide cycle?
8. How do animals help plants in the oxygen and carbon-dioxide cycle?
9. What are fossil fuels?
10. Water, air, soil, forests, and fossil fuels are all natural resources. Name one way you use each one of these.

Mad Scientist Challenge: Habitats

What is your habitat? Is it your bedroom? The city you live in? On a separate sheet of paper draw a picture of your habitat, or one of them. In your picture, include the community of organisms that live in your habitat.

Chapter Thirteen

Unit 2 Review

Answer the following questions on a separate sheet of paper.

1. All organisms respond to their environment. Describe one way plants respond to the movement of the sun.

2. Name the main parts of a cell. Why is DNA important to a cell?

3. Scientists once grouped fungi in the plant kingdom. Why did they change their minds?

4. Are worms vertebrates or invertebrates? How about fishes? Humans? What is the difference between vertebrates and invertebrates?

5. Plants use chlorophyll, sunlight, water, and carbon dioxide to make their food. What is this process called?

6. Gregor Johann Mendel was a scientist who worked with plants. What did he discover?

7. What are fossils? What can scientists learn from fossils?

8. Organs are grouped together in the human body to form systems. Name three systems and describe what they do.

9. White blood cells help the body fight disease. How do they do this?

10. Why do scientists say that cigarettes are harmful? Which systems and organs in the body are affected by smoking cigarettes?

Physical Science

Unit 3

Chapter 14
Properties of Matter

Chapter 15
Energy and Change in Matter

Chapter 16
Force and Motion

Chapter 17
Work and Machines

Chapter 18
Heat, Light, and Sound

Chapter 19
Electricity and Magnetism

Chapter 20
Energy Resources

Chapter 21
Computer Technology

Chapter 14

Properties of Matter

This is the famous Hope diamond, which is on display at the Smithsonian in Washington, D.C. It may be hard to believe, but diamonds and coal are forms of the same element—carbon.

Chapter Learning Objectives
- Explain what an atom is and identify its parts.
- List five properties of matter.
- Name the three states of matter.

Words to Know

compound a substance that is made when two or more elements join chemically

chemistry the scientific study of what substances are made of and how they can change when they combine with other substances

density the amount of matter per unit volume

electrons particles with negative electrical charges, surrounding the nucleus of an atom

gas matter without a definite shape or a definite volume

liquid matter with a definite volume but no definite shape

mass the amount of matter in something

mixture a substance that is made when two or more elements mix but do not join chemically

neutrons particles with no electrical charge, found within the nucleus of an atom

physics the scientific study of energy and how it interacts with matter

properties qualities of matter such as color, shape, odor, and hardness

protons particles with positive electrical charges, found within the nucleus of an atom

solid matter with a definite shape and volume

solution a kind of mixture in which one substance is dissolved into another

What do you have in common with a sandwich? How about beach sand, race cars, soda, pigs, or the air you breathe? Not much, you might say. Well, you have one important thing in common with all these things.

Along with everything on Earth that takes up space, you are made of matter.

You have already learned some things about matter. You know that it is made of tiny units called atoms. And you know that atoms can join together to make molecules.

In this chapter you will review what you already know about matter. You will also learn about several properties of matter. **Properties** are qualities that describe a thing.

Physical Science

The scientific study of matter and energy is called *physical science.* Some physical scientists concentrate on understanding matter. They study what matter is made of and how it can change. This field of study is called **chemistry**. Other scientists concentrate on understanding energy. They try to understand what energy is and how it interacts with matter. This field of study is called **physics**.

Chemists must understand a lot about physics, however. And physicists must understand chemistry, too. As you read on, you will see that the study of matter and energy go hand in hand.

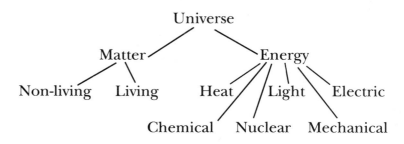

Chemists study matter and how it can change. Physicists study energy and how it interacts with matter.

Elements

You have already learned that all matter is made of elements. There are 103 known elements on Earth. Gold, silver, helium, oxygen, and nitrogen are examples of elements.

Elements are substances that cannot be broken down into simpler substances. Chemists use symbols to write about elements. A symbol is a shorthand way for writing a name. The chart on this page lists a number of the elements and their symbols. Look at Appendix D for the symbols for all of the elements.

Element	Symbol
Aluminum	Al
Arsenic	As
Calcium	Ca
Carbon	C
Chlorine	Cl
Chromium	Cr
Cobalt	Co
Copper	Cu
Fluoride	F
Gold	Au
Helium	He
Iron	Fe
Lead	Pb
Mercury	Hg
Platinum	Pt
Silver	Ag
Sulfur	S
Tin	Sn
Zinc	Zn

Amazing Science Fact

All substances made of the same element don't necessarily have the same properties. A diamond is one form of the element carbon. It is the hardest natural substance known. Diamonds are used to make cutting tools, record player needles, and of course jewelry.

Graphite is another form of the same element, carbon. But it is black, soft, and slippery. It is used to "grease" the moving parts in machines. It is also the part of pencils we often call the "lead."

The symbol for the element Hydrogen is H, and the symbol for Oxygen is O. A molecule of water is made of two hydrogen atoms and one oxygen atom. The symbol for this molecule is H_2O.

The Structure of Atoms

Elements are made up of tiny units called atoms. An atom has a central core called a nucleus. A cloud of tiny particles, called **electrons**, surround the nucleus. Different elements have different numbers of electrons. Hydrogen atoms have only one electron. The uranium atom has 92 electrons. Electrons move around the nucleus at very high speeds.

You probably remember that cells have nuclei, too. But they are very different than the nuclei of atoms. What do you think is the one thing that both nuclei have in common?

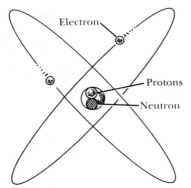

Atoms are made of electrons, protons, and neutrons.

Inside the nucleus of an atom are two types of particles. One is called a **neutron**. The other is a **proton**. Every atom has the same number of protons as electrons. The number of neutrons varies with the kind of atom.

Electrons and protons have opposite electrical charges. Opposite electrical charges attract each other. This is what holds electrons close to the nucleus of the atom. The electron is held by an attraction to the proton. Protons have positive electrical charges. Electrons have negative electrical charges. Neutrons don't have any electrical charge at all.

Science Practice
Answer the following questions on a separate sheet of paper.
1. What is chemistry?
2. How many electrons does a hydrogen atom have?
3. Do all substances made of the same element have exactly the same properties?
4. What are the two types of particles found inside the nucleus of an atom?

The Properties of Matter
Properties are qualities that make it possible to tell one kind of matter from another. A few properties of matter are color, shape, odor, and hardness. For example, mercury is silver and liquid. It does not have a noticeable smell.

Another property of matter is **mass**. Mass is the *amount* of matter in a thing. For example, a volleyball is about the same size as a bowling ball. But there is much more matter in a bowling ball than in a volleyball. It has a greater mass.

Density is another property of matter. Scientists define density as the *mass per unit volume*. By "per unit volume" they mean *for any given space*. So, density is the number of atoms (mass) in a given space. Think of a loaf of bread. Imagine balling it up and packing it together as tightly as possible. The balled up bread would take up less space than the original loaf. It would have a lesser volume than it did before you balled it up. Yet the amount of bread would be the same. So, its mass would be the same. The mass per volume, however, would be greater. The balled up bread would be *more dense*.

Since cork is less dense than water, it floats. The piece of lead is more dense than water so it sinks to the bottom.

Amazing Science Fact

You may have heard of "Fool's Gold." This is a substance called iron pyrite. It has many of the same properties as gold. For one thing, it looks just like gold. But a chemical test tells the truth. A few drops of acid will dissolve iron pyrite and give off a bad smell. True gold will not change form under "the acid test."

States of Matter

Matter has three different states: solid, liquid, and gas. For example, when H_2O is liquid it is water. In solid form H_2O is ice. In gas form H_2O is steam.

Solids have both definite shape and definite volume. Gold and silver both occur in nature as solid elements. If they are heated, they can be turned into liquids.

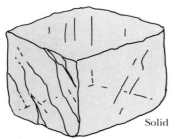

Solids have both definite volume and definite shape.

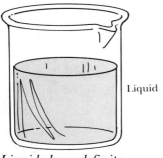

Liquids have definite volume. But they don't have definite shape.

Gasses have neither definite volume nor definite shape.

Like solids, **liquids** have definite volume. But they do not have definite shape. Think of a glass of water. If you poured that water into a bowl, the volume of water would still be the same. It would take up the same amount of space. But the shape of the water would change. It would take on the shape of the bowl. Now think about moving a solid, such as a plum, from a glass to a bowl. The plum would not change shape or volume.

Gas is also matter. That is, it takes up space. But it does not have definite shape or definite volume. A gas will spread out over a container of any size or shape. Think of the steam filling a small bathroom after a hot shower. The steam fills the whole bathroom. But suppose you took a shower in a much bigger bathroom. The same amount of steam would spread out over the bigger room. Air, hydrogen, helium, oxygen, and carbon-dioxide are all gasses.

Molecules in all substances are constantly in motion. But molecules in solids are packed very tightly together. They move very little. Molecules in liquids are spread further apart. They move a little more than the molecules in solids. The molecules in gases have even more room to move.

Compounds, Mixtures, and Solutions

You know that atoms can join together with other atoms to make molecules. The substance formed when two or more elements join together is called a **compound**. The elements in a compound have a *chemical bond*. In most chemical bonds, atoms share electrons. Water is a compound made of the two elements oxygen and hydrogen. Rust is a compound made by a chemical bond of iron and oxygen. Sugar, salt, and soap are also compounds.

A **mixture** is different from a compound. A mixture is two or more elements or compounds that are mixed together but not chemically joined. For example, soil is a mixture of rock, sand, and plant and animal matter. These parts of soil aren't joined chemically.

Try This Experiment
Put a small amount of salt water (no more than a quarter inch deep) in a wide bowl. Let it stand in a sunny place for several days. The water will evaporate. You will be left with salt.

This happens because salt and water do not bond chemically. The salt and the water stirred together make a mixture, not a compound.

A **solution** is a special kind of mixture. In a solution, one substance is dissolved into another. For example, the salt water you made in the above experiment was a solution.

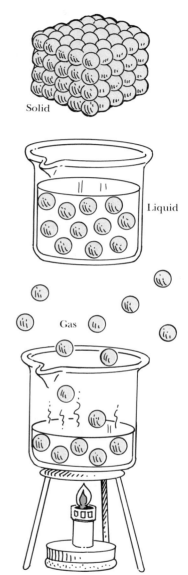

In a solid, atoms are packed close together. In a liquid, atoms are further apart than atoms in a solid. In a gas, atoms are further apart than atoms in a liquid.

Chapter Fourteen 159

Chapter Review

Chapter Summary

- Elements are substances that cannot be broken down into simpler substances. All elements are made of units called atoms. Each atom has a central core called a nucleus. Inside the nucleus of an atom are protons and neutrons. Moving around the nucleus at high speeds are the atom's electrons. Protons have positive electrical charges. Electrons have negative electrical charges. Neutrons have no electrical charge.

- The properties of matter include color, odor, size, shape, mass, and density. Mass is the amount of matter in something. Density is the amount of matter per unit volume.

- Matter can be solid, liquid, or gas. Solids have definite shape and volume. Liquids have definite volume, but not definite shape. Gas does not have definite volume or shape.

- When the atoms of two or more elements join, they make a compound. They are chemically bonded. When two or more elements mix, but do not bond chemically, they make a mixture. A solution is one kind of mixture. A solution is made by dissolving one substance into another.

Chapter Quiz

Write the answers to the following questions on a separate sheet of paper.
1. How many elements have scientists named?
2. A hydrogen atom has one electron. How many protons does it have?
3. Suppose you pour an ounce of water from a glass into a bowl. Does the water's mass change?
4. If you boil water, does the density of the water change?
5. Is the electrical charge of protons positive or negative? How about electrons? Neutrons?
6. Name two solid substances, two liquid substances, and two gas substances.
7. What happens to a gas when it is moved from a small container into a big container?
8. Are the molecules further apart in water, steam, or ice?
9. A salad is made of different vegetables thrown together. Is salad a mixture or a compound?
10. Is salt water a compound of salt and water? Why or why not?

Mad Scientist Challenge

Draw a model of an atom on a separate sheet of paper. Include the nucleus, protons, neutrons, and the cloud of electrons. Label all the parts. Include the electrical charges of the protons and electrons.

Chapter 15

Energy and Change in Matter

When this volcano erupted, liquid lava poured out of its vent. Soon after it settled, the lava hardened into stone. Why do you think the lava hardened?

Chapter Learning Objectives
- Explain the difference between potential and kinetic energy.
- Name five different forms of energy.
- Explain how heat energy changes matter.
- Describe the difference between physical and chemical changes.

Words to Know

chemical energy energy stored in molecules

condensation the process by which gas turns into a liquid

electrical energy energy produced and carried by the electrons in a substance

energy the ability to do work

heat energy energy produced by the motion of molecules

kinetic energy energy in movement

light energy energy produced by the motion of waves of light

mechanical energy energy produced by the moving parts of a machine

nuclear energy energy stored in the nucleus of an atom

nuclear fission the splitting of atomic nuclei, resulting in great energy release

nuclear fusion the joining of atomic nuclei, resulting in great energy release

potential energy stored energy

All the world is made of matter. And it takes energy to power all that matter. Without energy, rivers wouldn't flow. The Earth wouldn't go around the sun. You couldn't move a muscle. Nothing would happen at all.

In this chapter you will learn what energy is and how energy affects matter.

What Is Energy?

Energy has no mass. Yet it is as real as matter. There is energy in all things whether they are moving or still.

Energy is the ability to do work. In science, to work means to move something. There are two kinds of energy. **Potential energy** is stored energy. **Kinetic energy** is energy in motion.

For example, a running river has kinetic energy. Water held back by a dam has potential energy. A rock perched on the edge of a cliff has potential energy. But if that rock begins falling, the potential energy changes into kinetic energy.

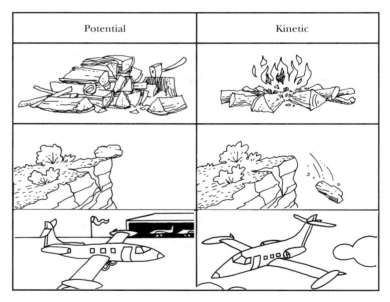

Potential and kinetic energy

Energy can be stored in matter. And it can be released from matter. A stick of wood, for example, stores energy. When it is burned, it releases energy. The energy it releases is in the form of heat and light.

Where do you get your energy? The energy our bodies need is stored in food. Before you eat your breakfast the food has potential energy. As you digest it, the energy stored in the food is released. As you walk to school, that energy turns into kinetic energy.

Energy that comes from the sun is called solar energy. It can be turned into heat energy or electrical energy. Many people use solar energy to heat their homes.

The Different Forms of Energy

There are several forms of energy. **Heat energy** is the energy an object gets from the movement of its molecules. The molecules in a hot object move faster than the molecules in a cold object. By rubbing your hands together you can increase the amount of heat energy in your palms. Notice how they become warmer.

Light energy is produced by the motion of waves of light. The main source of light energy comes from the sun.

Electrical energy is produced and carried by the electrons in a substance. Remember that individual electrons have negative electrical charges. A path of those negative charges is what carries electrical energy. Lightning is one example of electrical energy.

Lightning is electrical energy.

Chemical energy is energy stored in molecules. Molecules are made of atoms joined together by chemical bonds. When these bonds are broken, chemical energy is released. Cars use the chemical energy in gas. Humans use the chemical energy in food.

Mechanical energy is the energy released when the parts of a machine move. A windmill uses mechanical energy. So does a bicycle.

Bicycles use mechanical energy.

Chapter Fifteen   165

Nuclear energy is the energy stored in the nucleus of an atom. This energy is released when a large nucleus breaks apart. The breakup of a nucleus is called **nuclear fission**. Nuclear energy can also be released when two or more nuclei join together. This is called **nuclear fusion**. Nuclear power plants use nuclear energy to make electricity.

Energy can never be created or destroyed. But it can change form. Solar panels on a house change light energy into heat energy. Chemicals in a battery get turned into electrical energy. A water wheel turns the kinetic energy of a river into mechanical energy.

Nuclear power plants produce electricity from nuclear energy.

Science Practice

Answer the following questions on a separate sheet of paper.

1. What form of energy does a bicycle use?
2. What form of energy do plants use in photosynthesis?
3. When you stand in front of a furnace, what form of energy warms your body?
4. What form of energy does a TV use?

How Heat Energy Changes Matter

All changes in matter require some kind of energy. Heat energy causes the molecules in a substance to move faster and further apart. When enough heat is added, many solids will change to liquids. This is called *melting*. Ice is the solid form of water. When you add enough heat energy to ice, it melts.

As you add more heat to a liquid substance, the molecules move even faster. And they move further apart. With enough heat, molecules will begin to break away from the substance. The substance changes into a gas. When a liquid changes into a gas, the process is called *evaporation*.

Heat can also be taken away from matter. When heat is taken away from a gas, it turns back into a liquid. This is called **condensation**. If you take heat away from a liquid, it becomes a solid again. This is called *freezing*.

The dew you find on the grass in the morning is caused by condensation. The sun goes away at night. The air becomes much cooler. This causes water vapor (a gas) to change into water. You find it on your lawn and call it dew. As the day warms up, this dew will evaporate again.

Physical and Chemical Change in Matter

There are two kinds of changes in matter: physical and chemical.

A physical change affects only the state, shape, or size of matter. The chemical makeup of the substance stays the same. If you drop a plate and break it, for example, the shape has changed. But it is still a plate. No chemical change occurs. Crushing, tearing, and grinding are all examples of physical change. Freezing, melting, boiling, and condensation are also physical changes. Ice has the same chemical makeup as water. It just has less heat energy in it.

In a chemical change, however, a substance with new properties is produced. Compounds are always the result of a chemical change. For example, a kind of acid is formed when milk sours. The substance is no longer milk. Another example of chemical change is burning wood. Once burned, wood changes into ashes and some gases that go into the air.

What other discoveries can you think of that were the results of mistakes? What does this tell you about mistakes?

People in Science: Roy Plunkett

In 1938, the Du Pont company asked Roy Plunkett to make a refrigerant. That's a gas used in air conditioners and refrigerators. So Plunkett mixed up a bottle of gas chemicals that he thought might work. Then he put the bottle aside.

When Plunkett returned to the bottle, he found something very strange. Instead of a gas, there was a very slippery, white powder.

Plunkett didn't know it right away, but he'd created one of the first plastics. Today we know it as teflon. It took many more years for Du Pont to develop teflon into products, such as the nonstick pans people use today.

Roy Plunkett

Science Alert

Sometimes scientists claim they can do things they cannot really do. Several hundred years ago a group of scientists insisted that they could change ordinary substances into gold. These people were called *alchemists*.

An alchemist named James Price actually gave a demonstration. Before a group of people, he mixed a red powder with mercury. Indeed, he produced pure gold. Or so it seemed. The group was very impressed. The gold was even tested and proven to be real.

The Royal Society of England heard of Price's accomplishment. They invited him to demonstrate his discovery before their group. But Price kept making excuses for not appearing. Finally he invited the Royal Society to his home. Then, in front of the group, instead of making gold, Price swallowed a poison and died.

Price was a fraud. He had faked his first experiment. However, many other alchemists strongly believed in their work. Unfortunately, this work never did lead to important discoveries. For every breakthrough in science, there are always many more dead-end paths.

Chapter Review

Chapter Summary

- Energy is the ability to do work. Energy has no mass. But it is present in all things.
- Energy has two states. Potential energy is energy stored in things. Kinetic energy is energy in motion.
- Energy can never be destroyed or created. But it can change forms. Heat, light, electrical, chemical, nuclear, and mechanical energy are some of the forms.
- Heat can change a solid into a liquid. This is called melting. Heat can also change a liquid into a gas. This is called either evaporation or boiling. Heat causes the molecules in a substance to move around and spread out. By taking heat away from a gas, it can be turned into a liquid. This is called condensation. By taking heat away from a liquid, it can be turned into a solid. This is called freezing.
- Changes in matter can be either physical or chemical. A physical change affects only the state, shape, or size of matter. Chemical changes alter the chemical makeup of the substance.

Chapter Quiz

Answer the following questions on a separate sheet of paper.
1. Does energy have mass?
2. What are the two states of energy? Give an example of each.
3. Name five forms of energy. Give an example of each.
4. What kind of energy causes a chocolate bar to melt in the sun?
5. What happens to the water in wet clothes when they dry on a clothes line?
6. When you take heat away from a liquid, what happens?
7. What do all changes in matter require?
8. Describe one way to make a physical change in a wooden chair.
9. Describe one way to make a chemical change in a wooden chair.
10. What did the alchemists try to make?

Mad Scientist Challenge

Study the picture on this page. Name all the forms of energy you can find in the picture. In each case say whether it is potential energy or kinetic energy.

Chapter 16

Force and Motion

Gravity pulls this skier down the slopes. But eventually he will stop. Do you know why?

Chapter Learning Objectives
- Describe the effects of gravity, friction, and centripetal force.
- Explain the difference between weight and mass.
- Give three examples of inertia.

Words to Know

centripetal force   a force that causes objects to move in a curved path

force   any push or pull on an object

friction   a force that resists motion

gravity   the force of attraction between any two objects that have mass

inertia   the tendency to stay at rest or in motion unless acted on by a force

lubricants   substances that reduce friction between moving parts of machines

motion   a change in the position or place of an object

weight   the measure of the force of gravity on an object

Physics has a lot to teach us about everyday life. For example, understanding physics could make you a better athlete. Physics explains why you should follow through when hitting a baseball. Physics explains why it is harder to stop a big football player than a small one. Physics even explains why a speed skier wants the slickest possible skis. In this chapter you will learn about force and motion.

What Is Force?

Force is any push or pull on an object. When you throw a baseball, you push it away from you. When you drag a sled, you pull it behind you. Your body pushes on a chair as you sit in it. A dog on a leash pulls its owner behind him. These are all examples of force.

Gravity and Weight

All objects that have mass are attracted to each other. This attraction is a force called **gravity**. The more mass the objects have, the greater the force of gravity. The Earth is a very massive object. So, its gravitational force is very great. Gravity is the force that holds us on the ground. It pulls all matter toward the center of Earth.

Imagine the Earth without gravity. If you let go of an apple it wouldn't fall. It would just float in the air. In fact, you would also just float away. There would be nothing holding you to the ground.

The Earth is by far the biggest object in our environment. It exerts the greatest gravitational pull on us. But the moon also has an effect. The tides move in and out because of the moon's gravitational pull.

Amazing Science Fact

An object must have an extremely great mass for us to be affected by its force of gravity. The Earth has a mass of about 7 million trillion tons. Because the Earth is so massive, our attraction to it is also great. Nothing else on Earth is massive enough to affect us with its gravitational pull.

The strength of gravitational force depends on the mass of objects. Objects with less mass have less gravitational attraction. But gravitational force also depends on the distance between objects. The further away two things are from each other, the less gravitational pull they have.

Weight is the measure of the force of gravity on an object. The stronger the pull, the greater the weight. Weight is different from mass. *Mass* is the amount of matter in an object. An object's mass will stay the same wherever that object is. But the weight of an object can change. That is, the amount of gravitational force on an object can change. For example, the moon is less massive than the Earth. Suppose you were walking on the moon. There would be less gravitational pull on you than if you were on the Earth. And you would weigh less on the moon than you do on Earth.

A scale measures gravitational force.

Weight can also change because of the distance between two objects. In fact, the further you get from the center of the Earth, the less you weigh. You weigh slightly less on a mountain top than at sea level. This is because mountain tops are farther away from the center of the Earth. The pull of gravity is less. Therefore your weight would be less.

The force of gravity is also weaker near the equator than it is at the north and south poles. This is because the Earth bulges. It is bigger at the equator. Therefore, an object on the equator is farther from the Earth's center. It will weigh less than an object at one of the poles.

Amazing Science Fact

The World Trade Center Towers are the tallest buildings in New York. They are 1,350 feet tall. You can ride an elevator to the top of these buildings. At the top of the Towers you would be so high that your weight would change. Can you explain why?

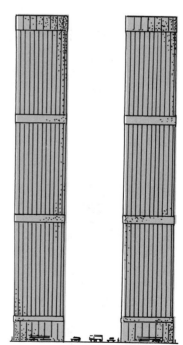

At the top of the World Trade Center you would weigh less than at the bottom.

Chapter Sixteen 175

On a flat surface, rolling friction brings a skateboarder to a stop.

Friction

Imagine an endless skateboard ride. You push off once, and you never have to push again. This would be possible if it weren't for **friction**. Friction is a force that slows or prevents motion. Three types of friction are rolling, sliding, and fluid.

When you drive a car, you must overcome rolling friction. The car wheels roll over the pavement. The pavement slows down the rolling of the wheels. The rougher the surface, the greater the friction. For example, ice is very slick. When a car rides on ice, there is much less friction than on dry pavement. This is why cars often have trouble stopping on ice.

When you slide across a polished floor in socks, sliding friction will stop you after a few feet. However, if you try to slide in sneakers, you will come to a stop very fast. There is greater sliding friction with sneakers than with socks.

When you row a boat, you must overcome the fluid friction in water. The boat pushes against the water. The water slows the boat's motion.

Friction explains why many things break down or wear out. The friction of a needle on a record will eventually wear off little pieces of the record. The friction of a bicycle chain on the cogs will wear down the metal. After a time, you will have to replace the cogs. **Lubricants** are substances that reduce friction between moving parts of machines. Oil, grease, and graphite are all lubricants. These substances are much more slippery than metal.

Engineers who design boats want to reduce friction. The more friction, the more energy a boat wastes. The fastest boats only touch the water a little bit. Less contact with the water means less water friction.

Try This Experiment

In the 1600s a man named Galileo studied falling objects. He found that light objects fall at the same speed as heavy objects. That is, unless air friction causes an object to fall more slowly than others.

Try Galileo's experiment. Drop a tennis ball and a baseball from the same height. Which hits the ground first? Then drop a feather and a stone. What happens now? Explain your results.

Centripetal Force

One law of physics says that all moving objects go in a straight path. That is, unless an outside force alters that path. **Centripetal force** is an example. It causes objects to move in a curved path.

Imagine you are driving a car fast around a curve. Your body gets pressed against the car door, right? Your body naturally wants to continue forward in the direction it was going. The car door is acting as a centripetal force keeping you in the path of the curve.

You can demonstrate centripetal force. Tie a ball to the end of a string. Hold onto the string and swing the ball in a circle. This is centripetal force in action. Holding the string creates the centripetal force that keeps the ball in its circular path. The track of an amusement park ride is also an example of centripetal force. The track keeps the ride car from going out in a straight line.

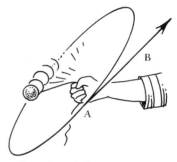

Centripetal force keeps the ball going in circles. If you cut the string at point A the ball would follow path B. It would go straight.

Centripetal force in action

Chapter Sixteen 177

Science Practice

Number a separate sheet of paper from 1 to 5. Match each word with its definition. Write a letter next to each number.

1. ____ gravity   a. any push or pull
2. ____ lubricant   b. prevents or slows motion
3. ____ friction   c. reduces friction
4. ____ force   d. the measure of gravity's pull
5. ____ weight   e. the attraction between any objects with mass

Motion and Inertia

Sir Isaac Newton was an English scientist. He was born in 1642, the same year Galileo died. Many people believe Newton was science's greatest thinker. He figured out many of the laws of gravity, including the law of inertia.

Inertia has to do with rest and motion. **Motion** is a change in the position or place of an object. Rest is a complete lack of motion. Objects tend to keep on doing what they were doing unless acted on by a force. This tendency is called inertia.

In other words, an object at rest will stay at rest forever unless a force moves it. Similarly, an object in motion will stay in motion forever unless a force stops it.

Imagine that you are a passenger standing in a crowded bus. The bus stops suddenly. Everyone is thrown forward. Before the bus stopped, the passengers were experiencing the inertia of moving.

Starting bus

Stopping bus

When the bus stopped, though, their bodies continued forward. After a second, the force that stopped the bus stopped the people, too.

Inertia is a property of all matter. The greater the mass of an object, the greater the inertia. A huge football player running downfield has greater inertia than a smaller player. It will take a stronger force to stop the bigger player.

Physics and Sports

Physics helps coaches and athletes master their games. Perhaps you have played baseball, tennis, or golf. In each of these sports, you exert force on a ball. Often, you want to hit the ball as hard as possible. You want to give it as much *force* as you can. In golf and baseball, a hard hit means the ball will go farther. In tennis, it means your opponent will have to move faster to return your shot.

The longer you exert a force on something, the greater that force will be. That's why a coach will tell you to "follow through" on your swing or stroke. By following through, you keep the bat, racquet, or club on the ball longer. You are exerting a force for a longer period of time. That makes the ball go faster and farther.

Think of a bullet fired straight up. Gravity and air friction will bring that bullet to a stop. Without these forces, however, the bullet would keep going forever.

Chapter Review

Chapter Summary

- A force is any push or pull on a thing. Gravity is the force that pulls all matter toward the center of the Earth. Weight is the measure of gravity. Mass is the amount of matter in an object.

- Friction is a force that resists motion. Sliding, rolling, and fluid are three kinds of friction. Lubricants are used to lessen friction between working parts of machines.

- All objects will move in a straight path unless affected by a centripetal force. Centripetal force causes objects to move in a curved path.

- Motion is a change in the position or place of an object. Every motion is caused by a force and can only be stopped by a force. Inertia is the tendency to stay at rest or in motion, unless acted on by a force.

Chapter Quiz

Answer the following questions on a separate sheet of paper.
1. Give one example of a pulling force.
2. Give one example of a pushing force.
3. What causes rivers to run downhill?
4. What force are people using when they put sand or gravel on icy roads in the winter?
5. Why do people put rough strips in their bathtubs?
6. What force keeps a person on a merry-go-round?
7. Wearing a seat belt protects you from possible harm by a certain force. Which force?
8. Explain the difference between weight and mass.
9. What force causes objects to move in a circular path?
10. What is inertia?

Forces on Cars

Below are three scenes involving cars. Tell which force or law of motion each scene is an example of. Write your answers on a separate sheet of paper.
1. A car falls off a cliff.
2. A car rolls to a stop.
3. A car moves slower and slower up a steep hill.

Mad Scientist Challenge: Baseball

Imagine that you are taking part in a baseball game. Name at least three forces or laws of motion that are a part of that game. Explain your answers on a separate sheet of paper.

Chapter 17

Work and Machines

This crane is a powerful machine. But it is based on a very simple idea. It uses a pulley to redistribute the weight of heavy objects.

Chapter Learning Objectives
- List the six simple machines.
- Explain "mechanical advantage."

Words to Know

effort force the force applied when doing work
fulcrum the support on which a lever turns
inclined plane a slanted surface used for raising objects from one level to another
lever a simple machine made of a bar that turns on a support
load the object moved through a distance in work
machine any device that can change the speed, direction, or amount of a force
mechanical advantage the number of times a machine multiplies an effort force
pulley a simple machine made of a wheel that turns on an axle. A rope or chain can be pulled around a groove in the wheel
resistance force the force that must be overcome in work
screw a simple machine made of an inclined plane wrapped around the length of a nail
wedge a simple machine made of a tapering piece of wood, metal, or other material
wheel and axle a simple machine made of a wheel attached to a shaft
work the force moving something through a distance

People and machines have a very close partnership. We use machines to do all kinds of work that would be impossible without them. Imagine trying to open a can with your bare hands. Imagine trying to cut the lawn without a lawn mower. Or trying to get from America to Europe without an airplane or a boat.

There are many many kinds of machines. Some are very complicated. An engine, for instance, is a very complicated machine. But all machines can be made from six simple machines.

In this chapter you will learn about these six simple machines. You will also learn how they help people work.

What Is a Machine?

In science, **work** means the force moving something through a distance. There are two forces involved in work. One is the force you apply. That is called the **effort force**. The other is the force that must be overcome. That is called the **resistance force** or the **load**. The load is the object you want moved.

Say you had to pick up a heavy sack of grain. Your lifting is the effort force. Gravity holding the bag down is the resistance force. Or say you want to drag the load across the ground. Your pulling is the effort force. Friction on the ground is the resistance force.

A **machine** is any device that can change the speed, direction, or amount of a force. The purpose of a machine is to make work easier. A machine's

When you lift something, the resistance force is gravity.

mechanical advantage is the number of times a machine multiplies an effort force. Mechanical advantage is a measure of how helpful a machine is. For example, say a machine has a mechanical advantage of 2. That means the machine doubles the effort force. If a machine has a mechanical advantage of 7 it multiplies the effort force by 7.

Simple machines are machines that produce work with one movement. There are six simple machines. Each one has some mechanical advantage. They are the lever, pulley, wheel and axle, inclined plane, screw, and wedge.

The six kinds of simple machines

Science is about asking questions. Sometimes you should ask questions about the scientists themselves.

Science Alert

In 1874, an American named John Worrell Keely announced a new discovery. He claimed to have built a *perpetual motion* machine. This is a machine that will run forever without needing outside energy. He claimed his machine could bend metal bars and fire bullets.

For twenty years, Keely lived on money that investors gave him to improve the machine. He claimed he would soon be able to have the machine run trains and do factory work. He raised more than a million dollars.

When Keely died, scientists searched his home. They found a pump under the floorboards. He'd used the pump to power his "perpetual motion machine." The machine was a fake! Today scientists know that perpetual motion machines are impossible to build. All machines must have energy put into them to work.

Levers

Crowbars, bottle openers, scissors, shovels, nutcrackers, and seesaws are all levers. A **lever** is any kind of bar or rod. It turns on a support called the **fulcrum**.

There are three classes of levers. In a class 1 lever, the fulcrum is between the effort force and the load. The load is the resistance force. A seesaw is an example of a class 1 lever.

In a class 2 lever, the load is between the effort and the fulcrum. A wheelbarrow is an example of a class 2 lever. The wheel of the wheelbarrow is the fulcrum. The wheel is the support on which the

This screwdriver is working as a lever. Do you know what class lever it is?

lever sits. The load, or resistance force, is in the wheelbarrow. And the effort force is in your arms. A nutcracker is another example of a class 2 lever.

In a class 3 lever, the effort force is between the load and the fulcrum. A broom is an example of a class 3 lever. The fulcrum is where your hands hold the broom. The resistance force is the floor. The effort is in the broom handle. Some other examples of class 3 levers are tongs, fishing poles, fly swatters, tweezers, and baseball bats.

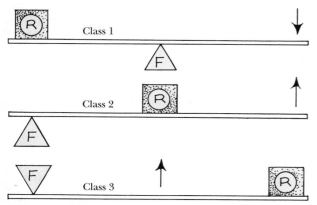

The three classes of levers

The Lever and Mechanical Advantage

Think of the bar of a lever in two parts: the resistance arm and the effort arm. The resistance arm is the part pushing or pulling the load. The effort arm is the part working to overcome that load.

By having a long effort arm, the work is spread over a greater distance. By increasing the distance over which the work is spread, you increase the mechanical advantage. So the longer the effort arm, the greater the mechanical advantage.

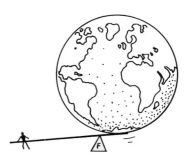

If you have a long enough lever and a strong enough fulcrum, you can lift anything.

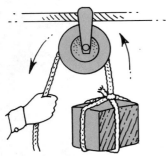

A pulley changes the direction of a load's force.

Pulleys

A **pulley** is a wheel that turns on an axle. Most pulleys are grooved. A rope or belt fits into the groove. The load is tied to one end of the rope. The effort force pulls on the other end.

Suppose you want to lift a bag of cement onto a platform. Lifting the bag by hand would be very difficult. You would be fighting gravity. Instead, you attach a rope to the load. Then you wind that rope around a pulley hanging above you. Now all you have to do is pull *down* on the free end of the rope. Of course you are still fighting gravity. But the gravity pulling the weight of your body will help you pull the rope. So the pulley changes the direction of the force to your advantage.

Some examples of pulleys are flag poles, clothes lines, water wells, and elevators. Sailors also use pulleys to hoist sails.

Wheel and Axle

When you turn the steering wheel of a car, you turn it only a few inches. But the car will make a turn of many feet. Car steering wheels are based on a machine called a **wheel and axle**. A wheel is fixed onto a rod, called an axle. As the wheel turns, so does the axle. The axle increases the force applied to the wheel.

Record players, egg beaters, and pencil sharpeners all use a wheel and axle.

A steering wheel is an example of a wheel and axle.

Try This Experiment

A doorknob is a wheel and axle. The knob is the wheel. The shaft of the door knob is the axle. Get permission to remove the knob from a door knob. Then try to turn the shaft with your fingers. Can you turn it enough to open the door? If you can, it will be very difficult. Now put the knob back on. Since it is so much bigger than the shaft, you can turn it easily. It turns the shaft for you.

Inclined Plane

Suppose someone gives you this choice. You can climb up the face of a cliff. Or you can climb up a gradual road. Which would you take?

An **inclined plane** is a slanted surface used for raising objects to higher places. Inclined planes make the work of going up easier. The work is spread out over a greater distance. Mountain roads and ramps are both inclined planes.

The inclined plane is an interesting type of machine. It has no moving parts!

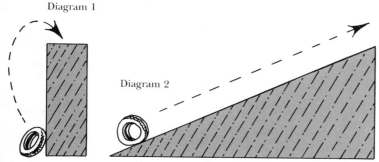

It would be very difficult to lift the car tire in diagram 1 to the top. It would be easier to roll it up the inclined plane in diagram 2.

Science Practice

Answer the following questions on a separate sheet of paper.

1. Name at least two examples of levers in your classroom.
2. Name at least one example of a pulley on your school grounds.
3. Name at least two examples of a wheel and axle on your school grounds.
4. When using an inclined plane, you must overcome friction. Suppose you must move a couch up a hill to a house. Would you push it up the grassy hill? Or would you push it up the paved driveway? Why? Some furniture movers use wooden platforms with wheels. How would this help?

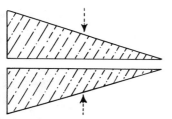

A wedge is made of two inclined planes placed back to back.

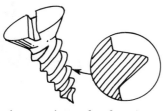

A screw is made of an inclined plane wrapped around a nail.

A **wedge** is made from an inclined plane. In fact, it is two inclined planes placed back to back. The wedge is thicker at one edge than at the other. Wedges are used to pry things apart. Knives, axes, needles, can openers, and razor blades are all wedges.

Screws are also made of inclined planes. A screw is an inclined plane wrapped around a nail. Screws are very efficient machines. Their mechanical advantage is high. When you turn a screw, a small effort force overcomes a large resistance force.

The First Machines

No one knows who first discovered simple machines. But they have been in use for at least 5,000 years. The Great Pyramid in Egypt was made of more than two million stone blocks. These stone blocks weigh more than two tons each. Some of them are 18 feet wide and more than 7 feet high.

How did people move these blocks without trucks and cranes? They must have used giant levers to lift the blocks. Surely they built long inclined planes to the top of the pyramid. Do you think they used ropes to drag the blocks up the inclined planes? Do you think they had a pulley system? No one knows for sure.

Compound Machines

A compound machine is made of two or more simple machines. Think of a mechanical pencil sharpener. The handle is a wheel and axle. The blades that cut the pencil are wedges. Another example is the can opener. A can opener has a lever to force the blade into the can. The blade itself is a wedge. And it has a wheel and axle to turn the opener around the rim of the can.

Record players, cars, typewriters, and lawn mowers are all compound machines.

What types of machines are combined to make a bicycle?

Granville T. Woods

People in Science: Granville T. Woods

Granville T. Woods was a 19th Century African-American inventor. He was born in Columbus, Ohio on April 23, 1856 and attended school until he was ten years old. His first job was in a machine shop. Later he worked on a railroad and in a mill. On all his jobs, he studied the machines. When he was 22, he got a job as an engineer on a steamship. Within two years, he became the Chief Engineer.

Soon after that, Woods opened his own machine shop. He began making electrical inventions. He sold his communications invention to Bell Telephone System and his electric railway idea to General Electric Westinghouse Company and Thomas Edison also bought some of Wood's inventions.

On a few occasions, Thomas Edison and Granville Woods met in court to argue over who had rights to certain electrical inventions. Both times, Woods won. Finally, Thomas Edison offered Woods a job with his firm. But Woods chose to work on his own. He sold most of his inventions through his company, Woods Electric Company.

On the Cutting Edge

By placing computers in machines, scientists can get both "thinking" and "muscle work" done for them. These "thinking" machines are called robots.

You may have heard of land rovers, tough trucks that can travel over rugged land. Space rovers are new traveling robots made to roam the surface of Mars.

Space rovers carry scientific instruments. These instruments take pictures of Mars, as well as study the soil, air, and weather. Two of these robots are scheduled to go on a 1996 mission to Mars.

One space rover, called Rocky 3, is the size of a toy, just 30 cm (one foot) long. It was built to be a model. But scientists now think it will be used just as it is for exploring Mars.

Another new robot, an eight-legged one named Dane, is going to travel into an active volcano in Antarctica.

Chapter Review

Chapter Summary

- Work is force moving something through a distance. A machine makes work easier by changing the speed, direction, or amount of a force.

- There are two forces involved in work. Effort force is the force applied to do the work. Resistance force is the force that must be overcome if the work is to be done. The load is the object that work moves.

- Mechanical advantage is the number of times a machine multiplies an effort force. Mechanical advantage is one way to measure how helpful a machine is.

- There are six simple machines. They are the lever, pulley, wheel and axle, inclined plane, screw, and wedge.

- Compound machines are made of two or more simple machines put together.

Chapter Quiz

Answer the following questions on a separate sheet of paper.

1. What is the force that must be overcome in work?
2. What force is put into a machine to do work?
3. If a machine doubles an effort force, what is the machine's mechanical advantage?
4. Give two examples of class 1 levers.
5. Give two examples of class 2 levers.
6. Give two examples of class 3 levers.
7. What kind of machine is a doorknob?
8. What kind of simple machine helps people in wheelchairs get into buildings and onto sidewalks?
9. Name a compound machine that is made of a wedge and an inclined plane.
10. How does a pulley change the direction of effort?

Matching Machines and Jobs

Number a separate sheet of paper from 1 to 5. Match each job with the machine best suited to it. Write a letter next to each number.

1. ____ cutting grass
2. ____ connecting pieces of wood
3. ____ writing letters
4. ____ making milk shakes
5. ____ cleaning rugs

a. typewriter
b. a screw
c. vacuum cleaner
d. lawn mower
e. blender

Mad Scientist Challenge: Invent a Machine

Think of a job you hate. Now invent a machine that will make that job easier. You can draw a picture of the machine. Or you can write a description of the machine. Use a separate sheet of paper.

Chapter 18

Heat, Light, and Sound

This huge antenna is actually a type of telescope. It is used by astronomers to study the universe. It picks up energy from outer space in the form of radiation.

Chapter Learning Objectives
- List the three ways that heat moves.
- Identify the parts of an energy wave.
- Explain why a red sweater looks red.
- Describe three properties of sound.

Words to Know

amplitude   the height of a wave

conduction   the bumping of molecules that moves heat through matter

convection   the transfer of heat within a gas or liquid by the movement of warmer particles

frequency   the number of wave cycles that pass through a point in one second

insulator   a material that does not conduct heat well

prism   a triangular-shaped object made of clear glass. It can break up a ray of white light into the colors of the rainbow.

radiation   energy that can move through a vacuum

reflection   light bouncing off an object

refraction   the bending of light rays when they pass from one material to another

spectrum   the rainbow-like band of color that is seen when white light is refracted

vacuum   the absence of matter

wavelength   the distance from the crest of one wave to the crest of the next

Hundreds of years ago American Indians hunted buffalo. To find the herds, they would press their ears to the ground. Why? Because sound travels better through solids than through gases. The hunters couldn't hear the stampede of hooves in the air. But sometimes they could hear it in the ground.

Have you ever noticed how clearly sound moves underwater in a swimming pool? Whales can hear each other from hundreds of miles away.

Using the science of sound, they found buffalo for food and clothing.

In this chapter you will learn how sound works. You will learn about heat and how it moves. You will also learn what light is and how it makes color.

Heat and Temperature

Heat is a form of energy. You feel a lot of different kinds of heat energy. You feel heat energy while standing in the sun. You feel it next to a fire. You feel your own heat energy when you sleep under blankets. There is some heat energy in everything, even the ocean. Here is why.

Remember that all matter is made of molecules in motion. The hotter the matter, the faster the motion of the molecules. This motion is *kinetic energy*. *Heat* is the measure of the *total* kinetic energy of the particles in a substance. *Temperature* is the *average* kinetic energy of the particles in a substance.

Imagine a cup of hot apple cider and a large jug of hot apple cider. They both have the same temperature. The average kinetic energy in both containers of apple cider is the same. But heat energy is the total kinetic energy of the particles in a substance. Since there are many more molecules of cider in the jug, its heat energy is greater.

This also explains why the Pacific Ocean has greater heat energy than a cup of hot coffee. The coffee has a higher temperature. But the ocean is so huge, that it has a tremendous amount of kinetic energy. Its total kinetic energy is much greater than that in the cup of hot coffee.

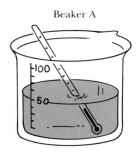

Beaker A

Beaker B

There is twice as much heat energy in beaker B as in beaker A.

Heat always moves from a warmer place to a cooler place. That is why you become cold when you step outdoors on a cold day. The warmth moves out from your body into the cold air. That is also why you can heat up a piece of cold pizza in the oven. The oven's warmth moves into the cold pizza.

Conduction: How Heat Moves Through Solids
Heat can move from one place to another in one of three ways: conduction, convection, or radiation.

In **conduction**, heat is moved along by molecules bumping into one another. Suppose you put a spoon in a cup of hot coffee. The top of the spoon doesn't touch the coffee at all. Yet it will soon be hot. This is because of conduction. The coffee heats the bottom of the spoon. The hot molecules on the bottom of the spoon bump into other spoon molecules. This warms them up. In turn, these molecules warm up the next ones up the spoon. The heat moves right on up to the tip of the spoon.

Heat from the tea moves up the spoon by conduction.

Some kinds of matter are better heat conductors than others. A good *conductor* is a substance that moves heat quickly. Metals, especially copper, are very good conductors. Some kinds of matter are very poor conductors. Heat moves through poor conductors very slowly. Paper, wood, rubber, glass, and plastic do not conduct heat well. Poor conductors are called **insulators**.

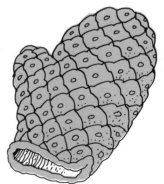

Do you think this pot holder is an insulator or a conductor?

There are many examples of conduction and insulation around your home. Many kitchen pans have copper bottoms. Why does copper make a good pan bottom? Pans also have plastic handles. Why is plastic a good material for pan handles? People insulate their homes to keep the warm air in. Can you think of some good materials to use for insulation? Would copper be good for insulation?

Convection: How Heat Moves Through Liquids and Gases

In conduction, heat moves along by molecules bumping into each other. In **convection**, heat moves along with the mass movement of particles.

Think of a floor heater in your home. The heater warms the air around it. The kinetic energy in this air becomes greater. As the molecules move faster, they spread out. In other words, the air becomes *less dense*. The light air then rises to the ceiling. The warm air pushes the cooler air down toward the heater. In turn, this cooler air gets heated and rises as well. A circular movement of air, called a *convection current*, is set in motion.

Convection works best in gases and liquids. Most weather is caused by convection currents in the air around Earth. There are many convection currents in the oceans. The Gulf Stream in the Atlantic Ocean is a convection current.

Convection: warm water rises and pushes cool water down.

Look back at page 157 for an explanation of density.

Radiation: How Heat Moves Through Space

Both conduction and convection move energy through matter. But parts of space have no matter. Where there is an absence of matter, there is a **vacuum**. Without matter, neither conduction nor convection will work. So, how do we feel heat energy from the sun?

Unlike heat energy, **radiation** can travel through a vacuum. The energy from the sun travels to the Earth as radiation or sunlight. Much of the radiation from the sun changes into heat energy as it strikes the Earth's surface.

Many Forms of Energy Travel in Waves

Energy waves carry energy from one place to another. They move in repeated patterns of motion. If you could see energy waves, they would look something like ocean waves. TV and radio messages also move through space in waves. Light and sound move in waves, too.

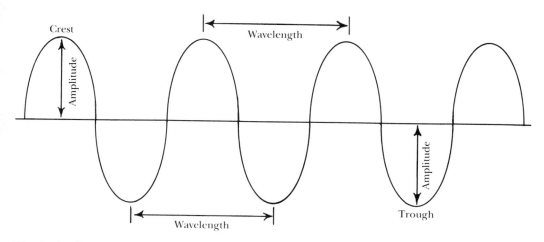

The basic characteristics of a wave: amplitude, wavelength, crests, and troughs

The top of a wave is called the *crest*. The bottom of the wave is called the *trough*. A **wavelength** is the distance from the crest of one wave to the crest of the next wave. The height of a wave is called its **amplitude**. The greater a wave's amplitude, the more energy it has. So a sound wave with a big amplitude will make a loud sound. A light wave with a big amplitude will make strong or bright light. The **frequency** of a wave is the number of wave cycles that pass through a point in one second.

Science Practice

Answer the following questions on a separate sheet of paper.
1. When you heat a pan of water, which kind of heat movement occurs in the water?
2. When the handle of a fireplace poker becomes very hot, which kind of heat movement warmed it?
3. When you sit in the sun, which kind of heat energy warms you?
4. Which kind of heat movement is used in a heated swimming pool?
5. When you put a hot water bottle in your bed, which kind of heat movement warms you?

What Is Light?

Light is a form of energy that travels in waves. Light waves always travel in straight lines. These straight lines of light are called *rays*. When a group of rays travel in the same direction, they become a *beam*.

Light travels through a vacuum at 186,282 miles per second. But air slows down light waves. Liquids slow them even more, and solids still more.

Light travels in many different wavelengths. We can only see a few of these wavelengths of light. What we can see is called *visible light*. Some light waves are too short for us to see. These short light waves are called *ultraviolet light*. *Infrared* light is made of waves that are too long for us to see.

On the Cutting Edge

A *laser* is a device that produces a very narrow, intense beam of light. In Chapter 1 you learned that lasers are used in compact disk players. But every day, scientists are finding more and more ways to use lasers. For example, doctors can use lasers, instead of knives, to cut out diseased tissue. Lasers make very clean cuts. A laser can also clear out a clogged artery.

Lasers are also used in industry. They can drill very tiny neat holes in very hard substances, such as steel. Lasers are used at check-out counters in grocery stores. They read the prices on products.

Some grocery store check-out counters have lasers. The lasers read prices from bar codes like this one.

What Happens When Light Strikes an Object?

Light can do four things when it strikes an object. It can pass straight through the object. It can be bent by the object. It can get absorbed into the object. Or it can bounce off the object.

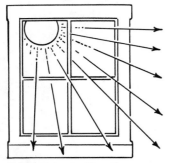

Light passes directly through glass.

Light passes through air, clear glass, clear plastic, and colorless liquids such as water.

Objects that are not clear absorb most of the light that hits them. The light passes into them but not through them. Light energy that is absorbed is changed to heat energy. Dark colors and rough surfaces are good light absorbers.

When light bounces off an object, it is called **reflection**. All materials reflect some light. Mirrors and objects with shiny surfaces reflect almost all the light that strikes them. The light reflecting off objects delivers images to your eyes. That is how you see.

If you stick a pole into clear water, it appears to bend. It looks like it bends just at the point where it enters the water. Of course the water doesn't really bend the pole. The water bends the light rays. This bending of light rays is called **refraction**.

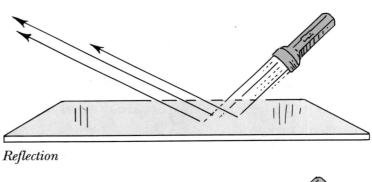

Reflection

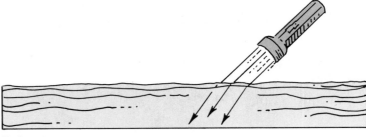

Refraction

Refraction is caused by a change in the speed of light as it passes from one substance to another. As light strikes the water it slows down and bends away from the surface of the water.

Where Does Color Come From?

Color is determined by the wavelength of light. Visible light is made up of several different wavelengths. These different wavelengths make bands of colors. Red, orange, yellow, green, blue, indigo, and violet are the seven colors in visible light.

These are the same colors you see in a rainbow. Rainbows are caused by sunlight passing through drops of water. The drops of water refract the white light. Different colors of light bend different amounts, so they separate into a band of seven different colors. This band is called a **spectrum**. It is made of red, orange, yellow, green, blue, indigo, and violet. All these colors together produce white light.

You can create a spectrum yourself by letting sunlight shine through a prism. A **prism** is a triangular-shaped object made of clear glass.

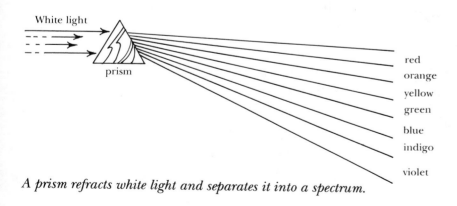

A prism refracts white light and separates it into a spectrum.

You know that objects absorb light, reflect light, or let light pass through them. The color you see is light reflected off an object. For example, imagine that you are looking at a red sweater. Light strikes the sweater. The sweater absorbs all the colors in visible light except red. It absorbs orange, yellow, green, blue, indigo, and violet. But the red part of light bounces off the sweater and enters your eyes.

A pure black object absorbs all the light hitting it. A pure white object reflects all the light striking it.

What Is Sound?

Sound is another form of energy that moves in waves. Sound waves travel about a million times slower than light waves. All sound is made by the vibration of matter. When you shout across a field at a friend, your vocal cords vibrate. The sound waves travel through the air. They make air molecules vibrate. Finally, they reach your friend's ear. There the waves cause tiny bones in your friend's ear to vibrate.

Try humming while holding your fingers at your throat. You can feel the vibration of your vocal cords. All musical instruments work by making vibrations.

Unlike light, sound cannot travel in a vacuum. Also, sound travels best through solids. It moves the least well through gas.

The loudness of sound is controlled by the amplitude of the sound waves. If you hit a drum hard, you will cause it to vibrate in big waves. You will make a loud sound. If you just tap the drum, you will cause only tiny waves. You will make a quieter sound.

The Speed of Sound in Various Substances

Substance	Speed (feet per second)
Air at 32° F	1,085
Air at 70° F	1,129
Aluminum	16,000
Brick	11,900
Glass	14,900
Quartz	18,000
Seawater at 77°F	5,023
Steel	17,100
Wood (maple)	13,480

Sound travels at different speeds through different substances.

Amazing Science Fact

Have you ever been in a thunder and lightning storm? You always see the lightning before you hear the thunder. The thunder is actually the sound of the lightning flash. You see the lightning first because light waves travel faster than sound waves. You can even use this information to measure how far away the storm is. Note the time between the lightning and the thunder. The greater the time between the two, the farther away the storm.

Try This Experiment

If your school has some musical instruments, ask permission to use two or three. Study the instruments. Find out which part or parts vibrate to make sound. Draw a picture of each instrument. Place an X on the spot where the sound begins. Then draw arrows to show the path of the vibrating parts. Include a person in the drawing. Show how the sound reaches the person's ears.

Chapter Review

Chapter Summary

- Heat is the measure of the total kinetic energy of the particles in a substance. Temperature is the average kinetic energy of the particles in a substance. Heat always moves from a warmer place to a cooler place.

- Heat moves through solids by conduction. Heat moves through liquids and gases by convection. Heat moves through a vacuum as radiation.

- Light and sound both travel in waves. Light can travel in a vacuum. Sound requires matter to travel. Sound is made by the vibration of matter.

- Objects that are not clear absorb most of the light that hits them. The wavelength of the light that reflects off objects determines their color.

Chapter Quiz

Answer the following questions on a separate sheet of paper.

1. Which has more heat energy, an icy mountain or a plate of hot stew?
2. What is conduction?
3. Name two materials that are good heat conductors.
4. Name two materials that are good insulators.
5. In a convection current, why does hot air or hot water rise?
6. How does light travel through space?
7. What three things can light do when it strikes an object?
8. Is the color you see on an apple passing through the apple? Is it getting absorbed into the apple? Or is it reflecting off the apple?
9. What causes sound when you pluck a string on a guitar?
10. Which travels faster, sound waves or light waves?

Draw a Light Wave

On a separate sheet of paper, draw a light wave. Label a crest, trough, wavelength, and amplitude.

Chapter 19

Electricity and Magnetism

Hundreds of years ago, no one knew for sure what caused lightning. Some said it was an angry message from the gods. Now we have a scientific explanation. Do you know what it is?

Chapter Learning Objectives
- Explain the relationship between electrons and electricity.
- Describe an electrical circuit.
- Name two properties of magnets.

Words to Know

battery   a device that changes chemical energy into electrical energy

circuit   an unbroken path along which an electrical current flows

conductor   a material through which electricity travels well

discharge   the letting go of extra electrons

electrical insulator   a material through which electricity does not travel

fuse   a weak link in an electrical circuit that is designed to break the circuit if it gets too hot

generator   a machine that turns mechanical or heat energy into electrical energy

magnet   a stone, piece of metal, or any solid substance that attracts iron or steel

magnetic field   the area around a magnet that exerts a magnetic force

static electricity   the electricity created when objects with opposite charges are attracted to each other

---

Imagine living without radio, TV, hair dryers, electric washing machines, light bulbs, or telephones. Well, only a short time ago people didn't have any of these things. All these tools run on electricity. And until about 100 years ago people didn't know how to control electricity.

In this chapter you will learn what electricity is. You will learn how it works. You will also learn about magnetism.

Electricity is not a recent invention. It has always existed in nature. What is recent is our ability to control electricity.

Electric Charge

Have you ever heard crackling as you brushed your hair? Or felt a small shock when you touched something metal? You have probably noticed how clothes that have been in the dryer stick together. Sometimes they make a crackling sound when you pull them apart. All these things are caused by electricity.

In Chapter 14 you learned that every atom has a nucleus. Inside the nucleus are positively charged protons. Circling the nucleus are negatively charged electrons. Most of the time atoms have the same number of protons as electrons. These atoms are called *neutral.* Neutral atoms have no charge at all.

But electrons can be rubbed off an atom. For example, suppose you rub a balloon on a wool sweater. Electrons from the atoms in the sweater rub onto the atoms in the balloon. Now the sweater has more protons than electrons. The sweater has a positive charge. The balloon has more electrons than protons. It has a negative charge.

Now suppose that the wall has a slight positive charge. If you hold the balloon to the wall it will stick there. The negative charge in the balloon will be attracted to the wall's positive charge.

The loss or gain of electrons makes an electric charge. Electricity is nothing more than the movement of electrons.

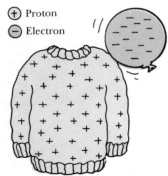

Electrons are rubbed off the sweater and onto the balloon. The balloon becomes negatively charged.

Static Electricity

Negatively charged objects have extra electrons. When two objects are both negatively charged they *repel* one another. This means they push each other away or try to move apart. Objects that have too few electrons are positively charged. Two positively charged objects will also repel each other. But objects that have opposite charges *attract* each other. This is what causes your socks to stick to your jeans in the dryer. Static electricity holds them together. **Static electricity** is the electricity created when objects with opposite charges are attracted to each other.

Try This Experiment

Rip a piece of paper into many small pieces. Then rub a comb vigorously on your pants leg or your shirt sleeve. Immediately hold the comb over the pieces of paper. The comb will attract the paper. You have created an electrical force in the comb.

Most charged objects don't keep their charges for long. Negatively charged objects give up their extra electrons. Positively charged objects take on needed electrons to become neutral again.

Think again of your laundry. If you pulled the socks and jeans apart in the dark, you might see little sparks. You might hear a crackling noise. The sparks and noise are the **discharge** of static electricity. The socks and jeans are returning to their neutral states.

Science Alert

It is dangerous to work with electrical appliances while standing in water. If you stand in water you become a *grounding object*. Electrons can move freely from the source of electricity through you to the ground. Electricity will pass through your body as if you were part of an electrical circuit. You will get a shock.

Lightning Is a Discharge of Electrons

Lightning is a huge electrical discharge. It is like a giant spark between a cloud and the Earth. Lightning can also occur between two clouds with opposite charges.

First a cloud builds up a lot of electrons. It gets a large negative charge. Then the cloud becomes attracted to something that is positively charged, such as a tree top. Suddenly, the extra electrons in the cloud "jump" to the tree top. This creates intense light and heat. The light is the bolt of lightning. The heat warms the air, causing it to expand very quickly. This quick expansion causes a loud noise called *thunder*.

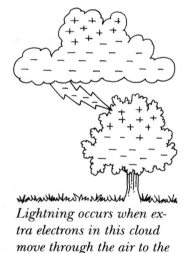

Lightning occurs when extra electrons in this cloud move through the air to the positively charged tree top.

People in Science: Benjamin Franklin

Benjamin Franklin was one of the greatest American politicians and scientists of the 1700s. He played an important role in the political formation of the United States. He was also one of the first scientists to experiment with electricity.

In 1752, Benjamin Franklin tied a key to the string of a kite. Franklin then flew his kite in the middle of a thunderstorm. A bolt of lightning struck his kite. Franklin put his hand near the key and felt it spark.

Franklin's experiment was dangerous. The charge from the lightning could have killed him. But he was happy. He had shown that lightning is an electrical discharge.

Benjamin Franklin showed that lightning is an electrical discharge.

Electrical Currents

You already learned that heat moves through some materials better than others. The same is true for electricity. Materials through which electricity travels well are called **conductors**. Metals are excellent conductors. Metal wire is often used to conduct electricity. But many other substances can conduct electricity, too. Even your body can conduct electricity.

Electrical insulators are materials through which electricity does not travel. Rubber and plastic are good electrical insulators. The rubber coating on the wire leading from the TV to the plug is an insulator. It protects you from the current running through the conductor inside.

Why do you think electrical wiring is always covered with plastic or rubber?

Chapter Nineteen 215

Science Practice

Write answers to the following questions on a separate sheet of paper.
1. List five ways you use electricity.
2. Which part of the atom moves in an electrical current?
3. Why do electrons "jump" to Earth in a thunderstorm?

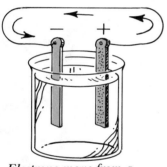

Electrons move from a positive charge to a negative charge.

To get a flow of electrons going through a conductor, people use batteries or generators. **Batteries** change chemical energy into electrical energy. Batteries are usually made of two different kinds of metal and some kind of acid. The acid removes electrons from one kind of metal and adds electrons to the other. So one piece of metal gets negatively charged and the other positively charged.

The two pieces of metal in a battery are connected by a conductor. Electrons in the positively charged piece of metal are attracted to the negatively charged piece. A current of electrons flows through the conductor to get to the negatively charged piece of metal.

Generators are sometimes used instead of batteries to start the flow of electrons. **Generators** are machines that make electrical energy from some other kind of energy. Some generators turn mechanical energy into electrical energy. Other generators turn heat energy into electrical energy.

Electrical Circuits

Many things in your house are powered by electricity. Lamps, dishwashers, hair dryers, stereos, and televisions are a few examples. The electricity to power these appliances must be controlled. You wouldn't want electrical charges shooting through your whole house!

The trick to making electricity useful is getting it to flow in paths. Electricians set up these paths for electrical currents to flow along. These paths are circular. That is, they come back to the place where they start and are not broken anywhere. An unbroken path along which an electrical current flows is called a **circuit**. A circuit includes the source of energy, such as a battery.

Look at this picture of a closed circuit between a battery and a light. The battery produces the source of the electricity that flows through the wire. As it passes through the light bulb, it heats a piece of material called a *filament*. The filament gets so hot it glows, producing light. The electricity then continues along the circuit and back into the battery.

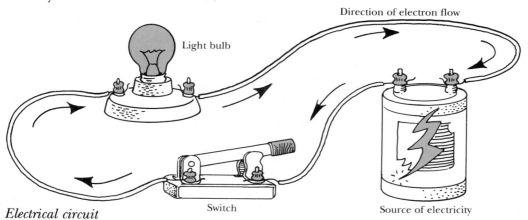

Electrical circuit

Some circuits have switches. If the switch is turned off, the circuit is *open* and the electricity cannot flow. If the switch is turned on, the circuit is *closed* and electricity can flow.

When the electrical energy heats the filament to make light, some of it is transformed into heat energy. And that heat energy is lost from the circuit. Sooner or later, the battery will run out of energy.

When too many appliances are put on one circuit, it can become *overloaded*. This means that too much electricity is running through the wire. An overloaded circuit can cause the wire to get too hot and start a fire.

Fuses are used to prevent overloaded circuits from causing fires. A **fuse** is a weak link in a circuit. It is made of metal wire that has a low melting point. When too much electricity flows through the wire, the fuse melts. And when the wire melts, the circuit is open and the flow of electricity stops.

When too much electricity flows through a fuse it melts, opening the circuit.

What Is Magnetism?

More than 2,000 years ago, the Greeks discovered stones with special properties. These stones attracted each other. They also attracted iron. These stones, called *lodestones*, are natural magnets. A **magnet** is any stone, piece of metal, or solid substance that attracts iron or steel.

You may use magnets to stick notes onto your refrigerator. Some people use them at their desk to hold all their paper clips together.

Magnetism is a property of all matter. But only a few elements have magnetism that is strong enough to be noticed. Iron, nickel, and cobalt are three elements with strong magnetism.

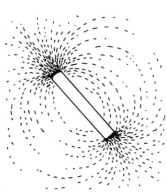

The iron filings sprinkled around this magnet show its magnetic field.

Magnetic Fields

Each end of a magnet is called a pole. If you hang a magnet on a string and let it swing freely, one end of the magnet will always point toward the north. The other end will always point south.

Magnetic poles act the same way electrical charges do. Poles that are the same are called *like* poles. Like poles repel each other. Opposite poles attract each other. The magnetic force is strongest at the poles. But the force is felt all around the magnet. The area in which magnetic force is active is called the **magnetic field**.

The electrons in atoms are like tiny magnets. But these electrons are spinning around the nucleus of an atom *randomly*. That is, they are not in any order. So their magnetic pulls are in all different directions. When a piece of metal becomes magnetized, the electrons all line up. The tiny fields of magnetism in electrons all turn so that the north poles all face north. Working together like this, the electrons create magnetic pull.

If a magnet is hammered or heated, it can lose its magnetism. Hammering or heating stirs up the tiny magnetic fields of the electrons. That makes the electrons return to random positions.

A compass is a small tool that helps people find north, south, east, and west. The needles on compasses are made of magnets that spin freely. The negative pole of the magnet points south. The positive pole points north.

Chapter Nineteen 219

Chapter Review

Chapter Summary

- Electricity is the movement of electrons from one object to another. When a substance has more electrons than protons it is negatively charged. When a substance has fewer electrons, it is positively charged.

- Objects with opposite charges are attracted to each other. This is static electricity. When a negatively charged object returns to neutral, it discharges the extra electrons. This is what causes lightning.

- With the help of a conductor, electricity can be made to flow in a path. This is called a current. Batteries or generators must be used to get the flow of electrons started.

- A closed current is called a circuit. People use electrical circuits to power appliances in their homes. If a circuit is broken, the flow of electricity stops.

- A magnet is a stone, piece of metal, or any solid substance that attracts iron or steel. Electrons are all tiny magnets. When they line up in a substance, they work together to make a strong magnet. If a magnet is hammered or heated, the electrons go back to random positions. The magnetism is broken.

Chapter Quiz

Write answers to the following questions on a separate sheet of paper.

1. What is static electricity?
2. If an object has extra electrons, does it have a negative or positive charge?
3. If an object has extra protons, does it have a negative or positive charge?
4. What is the difference between a battery and a generator?
5. What is an electrical circuit?
6. Is the human body a conductor or an insulator of electricity?
7. How could a fuse help prevent a fire in your home?
8. What is a magnet?
9. Name two ways you might use magnets or magnetism in your life?
10. How can a magnet lose its magnetism?

Mad Scientist Challenge: Electricity at Home

Some homes use gas instead of electricity for certain things. Look around your home. Copy this chart on a separate sheet of paper. Then put a check under *gas* or *electricity* for each item.

	Gas	Electricity
Cooking		
Heating		
Television		

Chapter 20

Energy Resources

These windmills use an old idea to make energy for the modern world. As the wind turns their blades, it makes mechanical energy. What do you think happens when the wind stops blowing?

Chapter Learning Objectives
- List and define five different sources of energy.
- Outline one benefit and one problem with each source of energy.

Words to Know

geothermal energy energy from hot water trapped under layers of rock deep in the Earth

geyser a hot spring from which steam and hot water shoot into the air

hydroelectric energy electrical energy produced by the movement of water

solar collector a tool that is used for collecting sunlight and transforming it into heat energy

turbine a machine driven by the force of a moving fluid

When a volcano explodes it releases a lot of energy. Ocean waves and the blowing wind are also full of energy. There's an almost unlimited supply of energy in the burning sun. Even tiny atoms hold huge amounts of energy. The question is, how can people tap into this energy and use it?

You already know a lot about energy. You know the forms it can take: heat, sound, light, electrical, chemical, and mechanical. In this chapter you will learn more about how these sources of energy can be used.

Fossil Fuel

In Chapter 13 you learned that coal, petroleum, and gas are called fossil fuels. Nearly 90 percent of all the energy used in the United States comes from burning fossil fuels. Fossil fuels are used to run our

Nearly 90 percent of the energy used in the United States comes from burning fossil fuels.

Many plastics are also made of fossil fuels. Can you believe that plastic and gasoline are both made of dead plants and dinosaurs?

cars, heat our homes, and provide power for many of our factories.

But there are two serious problems with burning fossil fuels for energy. First, there is only so much fossil fuel in existence. These fuels take millions of years to form. People cannot make more of them. So when they run out, they are gone for good. The second problem with fossil fuels is that they cause air pollution when they burn. The waste products from burning fossil fuels are very poisonous and bad for the environment.

People can help to lessen both of these problems. The less fossil fuel that is used, the longer the supply of fossil fuel will last. Gasoline is saved when people get around by taking the bus, riding a bike, or walking. These ways of getting around also help to keep the air cleaner. Insulating your home will mean using less energy to keep it warm.

Nuclear Energy

Scientists have learned how to split apart the nucleus of an atom. This process, called *nuclear fission*, releases a tremendous amount of energy.

Nuclear fission takes place in nuclear power plants. There, millions and millions of atomic nuclei are split every second. This lets off a great deal of heat. The heat is used to turn water into steam. The steam is then used to power generators. The generators turn the heat energy into electricity.

Nuclear fission has some serious problems. For one thing, the fission process makes wastes that are very dangerous. These wastes may remain dangerous for millions of years. No one is sure what should be done with them. If they are buried, they may pollute the water and soil. Accidents at nuclear power plants are also a problem. Harmful substances have leaked out of a few power plants. Many people think nuclear energy is not worth the risks.

Scientists are developing another kind of nuclear energy. It is called *nuclear fusion*. Nuclear fusion joins atoms together. Energy is released in this process just as it is when atoms are split. Nuclear fusion is what causes all the heat and light energy in the sun.

Nuclear fusion is a very clean source of energy. The wastes from fusion are oxygen, water, and other safe things. The problem is that scientists do not yet know how to control the nuclear fusion process. If they can solve this problem, nuclear fusion may one day be a good, safe, source of energy.

Nuclear fission is also used to make nuclear bombs explode.

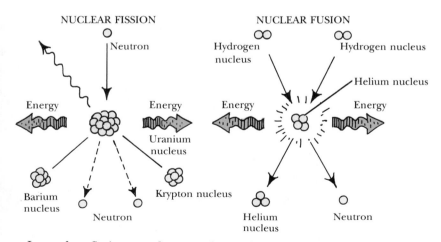

In nuclear fission, one large nucleus splits to form two smaller nuclei. In nuclear fusion, small nuclei join together to form a larger nucleus.

Chapter Twenty   225

Science Alert

The federal government is studying whether or not to store nuclear wastes inside a Nevada mountain. Yucca Mountain lies about 150 kilometers (about 100 miles) northwest of Las Vegas. The planned storage area would be built in rock, far above the water table. The nuclear wastes stored there would be dangerous for 10,000 years. People are concerned that flooding could cause the wastes to leak out of the mountain into our water. Earth scientists are studying the question. Do you think scientists can predict what will happen to a mountain 10,000 years in the future? Why or why not?

Science Practice

Answer the following questions on a separate sheet of paper.
1. What are fossil fuels made of?
2. Name three good ways to save fossil fuels.
3. What is the difference between nuclear fission and nuclear fusion?

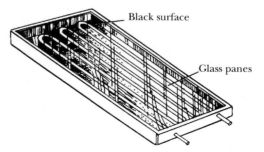

This solar collector is covered with glass. Inside, it has a black surface to absorb sunlight.

Solar Energy

Solar energy is energy given off by the sun. It reaches us mainly in the form of light energy. Imagine that you are sitting in a car in the sun. Pretty soon you start to get very warm. Sunlight hitting your car has been transformed into heat energy. The car has trapped the heat inside. This is also how a **solar collector** works.

Perhaps you have seen solar panels on buildings. Solar collectors or panels are usually dark in color. This dark surface absorbs the sunlight and transforms it into heat energy. The heat energy warms the water inside the solar collector. This water is circulated around the building for heat.

Solar energy can also be used to produce electricity. A device called a *solar battery* converts sunlight into electricity. But this is a new technology and the process is still very expensive.

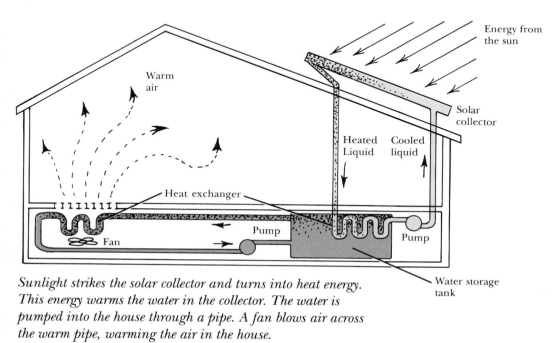

Sunlight strikes the solar collector and turns into heat energy. This energy warms the water in the collector. The water is pumped into the house through a pipe. A fan blows air across the warm pipe, warming the air in the house.

There are other problems with solar energy. Energy from the sun is powerful. But it is spread out over wide areas. Scientists must find ways to gather and store it. And they must find ways to do this inexpensively.

Energy from the sun is clean, plentiful, and free. Many scientists hope that solar energy will provide much of the world's energy in the future.

Hydroelectric Energy

Hydroelectric energy is produced by moving water. A dam is built to trap water. Water is allowed to come through the dam a little at a time. This water strikes the blades of a big **turbine**, causing them to turn. This starts an electric generator which turns the mechanical energy into electrical energy.

Some dams serve a double purpose. They create hydroelectric energy and store water for future use.

Dams provide energy to many cities in the United States. But people do not want to dam up too many rivers. Dams can hurt fish and other river life.

Geothermal Energy

Volcanoes prove that there are great stores of energy in the Earth. *Hot springs* and **geysers** also show this. Scientists have learned to tap this heat energy. It is called geothermal energy.

Geothermal energy is heat energy deep in the Earth. In certain places, there are large pockets of very hot water trapped beneath layers of rock. The steam from the hot water can be used to generate electricity.

The problem is, these pockets of heat are only found in a few places. Where there are no hot pockets, there is no geothermal energy.

Geysers shoot water high into the air. This water carries geothermal energy from deep within the Earth.

Wind and Tides

Wind has been used for energy for hundreds of years. All over the world, windmills pump water, grind grain, and make electrical energy. Of course windmills work best where there is lots of wind.

There is also a lot of energy in ocean waves and tides. In France there is a tidal power plant. This plant is able to make electric power from energy in the moving tides. This kind of energy is only useful on the coasts. And even there, conditions must be just right for setting up a plant.

Chapter Review

Chapter Summary

- Fossil fuels are the most widely-used source of energy. Fossil fuels are the remains of plants and animals that died millions of years ago. Fossil fuels are mined from the Earth. Unfortunately, there is only so much fossil fuel in existence. Sooner or later it will be used up.

- Nuclear fission is energy made by splitting atomic nuclei. This source of energy creates very dangerous wastes.

- Nuclear fusion is energy made by joining atomic nuclei. This is what causes all the energy in the sun. Scientists are still working on ways to control nuclear fusion.

- Solar energy is energy given off by the sun. This energy is clean, safe, and plentiful. But it is still very expensive to collect and store.

- Hydroelectric energy is made by moving water. Geothermal energy is tapped from pockets of hot water deep in the Earth. Windmills have been used for years to capture the energy in the wind. Some scientists are working on ways to get energy from ocean waves and tides.

Chapter Quiz

Answer the following questions on a separate sheet of paper.

1. Define fossil fuels and give two examples.
2. Which kinds of energy produce dangerous wastes?
3. Why are solar panels dark-colored?
4. Which kind of energy causes the heat and light energy in the sun?
5. What kind of energy is released when a volcano erupts?
6. What kind of energy does a dam help to make?
7. What kind of energy collection system would work well in a desert? Why?
8. What kind of energy production would not work in the middle of this country? Why not?
9. Where do windmills work best?
10. Which kind of energy do you use most?

Mad Scientist Challenge: Matching Energy Sources

Every kind of energy has at least one problem. On the left are energy sources. On the right are problems. Number a separate sheet of paper from 1 to 7. Match each energy source with a problem. Write a letter next to each number.

1. ___ fossil fuels
2. ___ nuclear fission
3. ___ solar
4. ___ hydroelectric
5. ___ geothermal
6. ___ wind
7. ___ nuclear fusion

a. creates dangerous wastes
b. still too costly and hard to store
c. can only be used on windy days
d. hot pockets are not everywhere
e. limited amount
f. damming rivers hurts wildlife
g. scientists do not yet know how to control it

Chapter 21

Computer Technology

This computer is used to help design cars. Why do you think a car designer might want to work on a computer instead of on drawing paper?

Chapter Learning Objectives
- Describe the main parts of a computer.
- Name six ways computers are used.

Words to Know

bit the smallest unit of information used by a computer; represented by a 0 or a 1

byte a string of eight bits standing for a single character

data information, facts, numbers, or letters processed by a computer

disk drive the machine that reads data on diskettes and records new data onto diskettes

diskettes plastic disks covered with magnetic material and used to store data; they can be removed from the computer

input data entered into the computer

keyboard the input device used to type information into the computer, much like a typewriter keyboard

magnetic tape plastic tape covered with magnetic material; used for recording data

monitor the part of the computer that has a screen showing what is going on in the computer

output processed data that comes out of the computer

processing working with data on a computer

program a set of coded instructions that leads a computer through certain tasks

A computer is a machine that performs calculations and processes information with astonishing speed and precision.

The most powerful computer can perform billions of calculations per second.

Computers can guide spaceships deep into space. They can help doctors figure out what kind of disease their patients have. Computers are used to predict the weather. Some computers can even talk and play music.

At first, computers may seem very complicated. But they are not as hard to use as you might think. Most computers made today are "user friendly." That means they are specially designed to be easy to use. Computers were developed to make people's work easier—not harder!

In this chapter, you will learn about the parts of computers. You will also learn how computers work.

The Parts of a Computer

You have probably seen computers. Most of them look like boxes with TVs sitting on top. The box is called the *main processing unit* of the computer. The processing unit is where the computer does its work. Computers process **data**, or information. **Processing** is the work the computer does with data. This might be sorting, calculating, listing, or some other kind of work.

The part of the computer that looks like a TV is called the **monitor**. The monitor shows what is happening inside the computer. If you type on the computer keyboard, the words you type appear on the monitor. If you are drawing, the picture appears on the monitor. Any messages from the computer also appear on the monitor's screen.

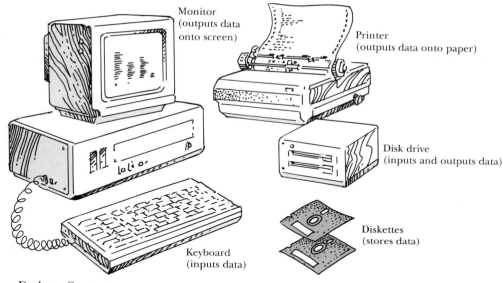

Desktop Computer

The monitor is an output device. **Output** is the processed information that comes out of the computer. An *output device* is any part of the computer through which that information comes out.

The **keyboard** is an example of an **input** device. An input device is any part of the computer that is used to enter information. The computer keyboard is a lot like a typewriter keyboard. But what you type on a computer keyboard goes into the computer instead of directly onto paper.

Eventually, people want a copy or a record of the data they've entered into the computer. If they want a record printed on paper, they use an output device called a *printer*. Data printed on paper is called *hard copy*.

Because computers process lots of information they must have a lot of memory. In the language of computers, "memory" is the place where the information is stored.

The largest computers fill whole rooms. The smallest ones fit inside wrist watches.

Chapter Twenty-One 235

This diskette can store about 100 pages of writing.

Diskettes are a lot like cassette tapes. Both are made of magnetic material. But instead of recording sounds, diskettes record data processed by a computer.

Computers work by sending electrical signals through wires.

There are different kinds of computer memory. Some memory is in the processing unit. But information can also be stored on diskettes or magnetic tape. **Diskettes** are plastic disks covered with magnetic material. **Magnetic tape** is plastic tape covered with magnetic material. Both store data for future use. But they can be removed from computers the way cassette tapes can be removed from tape players.

Diskettes can store the words of a novel. They can store the accounts of a bank. They can store mailing lists. Suppose a computer user wants to use information that's on a diskette. He or she simply puts the diskette into the computer. Computer users often collect whole libraries of diskettes with different kinds of information on them. Diskettes go into a special slot in the computer called a **disk drive**. The disk drive is like a tape player. It "reads" the information on the diskette and sends it to the processing unit. The disk drive can also copy new information onto a blank diskette.

How Computers Work

Computer programs are written in special codes. All computer codes use only two numbers: 0 and 1. The 0 stands for *off*. The 1 stands for *on*.

Each 1 or 0 is called a bit. A **bit** is the smallest unit of information used by a computer. A string of connected bits is called a **byte**. Each byte stands for a letter, number, or symbol. These bytes are strung together to make computer programs.

A **program** is a set of coded instructions that guides the computer through its tasks. Different jobs require different programs. Scientists call computer programs *software*. The computer itself, and all of its mechanical parts, are called *hardware*.

On the Cutting Edge

"Pen computers" or "smart paper" are new, tiny computers. About the size of a notebook, these smart computers can read handwriting. A user writes directly on the machine's screen with a special pen. The computer can read the letters. These computers are great for people who don't know how to type. But they aren't so great for people with poor handwriting. The writing has to be neat for the computer to track it.

Science Practice

Number a separate sheet of paper from 1 to 6. Match each part of a computer with the job it does. Write a letter next to each number.

Parts
1. ___ main processing unit
2. ___ monitor
3. ___ diskette
4. ___ keyboard
5. ___ printer
6. ___ disk drive

Jobs
a. makes paper copies or output
b. shows what's going on inside the computer
c. stores data
d. processes data
e. reads data on a diskette or puts new data onto a diskette
f. lets the user type in information

On the Cutting Edge

Sitting in a chair at home you can explore a coral reef in Australia. Or try to land a spaceship on Mars. Or become a dinosaur sloshing through a swamp thousands of years ago.

Cyberspace, a new computer technology, is a kind of artificial reality. The user puts on a headset that has two small computer screens, one in front of each eye. The screens feed the computer imagery to the brain through the eyes. The user also wears skintight gloves—sometimes even a skintight body suit. The gloves and body suit have electromagnetic sensors that track the user's movements. These movements are fed to the computer. To pick up a cyberspace object, you just have to make a fist. To fly in a certain direction, just point a finger.

The user feels as if he or she is moving through real space. The experience feels so real that people get seasick on sailing adventures. Or sore muscles from hard tasks. Or dizzy from dancing in circles.

Besides fun, there are some practical uses for cyberspace. Architects can use it to let a homeowner walk through a house before it has been built. Medical students can practice surgery. Pilots can practice flying.

The Many Uses of Computers

Computers have changed the way we work and do business. They have also changed the way we play. Computer games are a lot of fun—and sometimes they are educational, too.

Business people use computers to help them make decisions. A computer can process much more data than a person can, and a lot faster, too. Suppose a businesswoman were trying to decide whether to invest money in a piece of equipment. Her decision might be based on thousands of pieces of information. Using the right program, her computer could process all the pieces of information and help her make the right decision.

This woman is learning to drive on a computer. If she "crashes" on the computer, it is a lot less serious than if she crashes on the road.

In the United States computers are used to count votes during elections. Almost 100 million people vote in presidential elections. You can see how computers would save lots of time.

Computers are also good for making models of real-life situations. For example, some people learn to drive by using computers. The computer screen shows streets and other traffic. Without the danger of real roads, people can safely learn traffic rules. They can practice what to do in dangerous situations without actually being in any danger.

Libraries, banks, and most offices handle huge amounts of data. The data may be in the form of numbers, addresses, or book titles. Within seconds, a computer can bring up any combination of information that an employee needs.

Suppose you want to see if your library has any books about your favorite musician. You would type the musician's name onto the computer keyboard. In seconds, a message will appear on the monitor screen. This message may give you the names of books about the musician. It might even tell you the names of some records your library has. If your library has a computer, try it.

Computers are also very useful for writing. Working with words on a computer is called word processing. With a word-processing program writers can watch their words appear on the screen as they write. It is easy to make changes on a computer. That way writers can get their work just the way they want it. Then they can print out a perfect copy. Many word-processing programs even have dictionaries that check spelling automatically.

Computers are smart, but there are still a lot of things only people can do. Can you imagine a computer writing a beautiful love poem?

In the few years since small computers have become widely used, many computer jobs have been created. Data entry employees enter information into computers. The information is stored for later use. Other people work in factories where computers are made. Other workers help fix computers when they break down. And still others write the programs that make computers work.

More and more libraries are putting their card catalogs onto computers.

People in Science: Manuel Berriozábal

There are a lot of great jobs in computers. How can you get them? Study math now.

Since 1979, mathematician Manuel Berriozábal at the University of Texas has run a summer math program called TexPREP for middle school and high school students. He knows that scientists begin their training young. He must be right. All the kids in his summer programs graduate from high school. And 80 percent of them go to college. Over half of Berriozábal's students major in science.

People with careers in computers start young. And they start with math.

Manuel Berriozábal

Chapter Review

Chapter Summary

- A computer is made of a main processing unit, a monitor, and a keyboard. The processing unit stores the information and does most of the work. The monitor is an output device. It shows the user what is going on in the computer. The keyboard is an input device. It is used to enter information.

- Diskettes are used to store information. A printer is an output device. It prints information in the computer onto paper.

- Computers follow instructions called programs. Programs are written in special computer codes using only the numbers 0 and 1. Every piece of information in a computer is stored as a combination of 0s and 1s.

- Computers have many uses. They are used to keep track of numbers. They are used to process huge amounts of data. They can be used for writing. And they are used to make models of real-life situations.

Chapter Quiz

Answer these questions on a separate sheet of paper.
1. What is processing?
2. Which part of the computer does the processing?
3. Where is information inside the computer stored?
4. Which part of the computer shows you what is going on inside?
5. Which part of the computer is used to enter information?
6. How is information stored outside the computer?
7. Where do you put a diskette so that a computer can "read" it?
8. What is the coded list of instructions that guides the computer through its tasks called?
9. Which two numbers are used in making computer codes?
10. What is a computer chip?

Mad Scientist Challenge: Charting Computer Uses

Copy this chart on a separate sheet of paper. Suggest at least one job the computer can do in each place.

Place	Computer Uses
office	
home	
school	
library	
music studio	

Unit 3 Review

Answer the following questions on a separate sheet of paper.

1. Explain the difference between mass and density.
2. Name and describe the parts of the atom.
3. Name six different forms of energy. Give an example of each form you named.
4. Does energy have mass?
5. What is centripetal force?
6. What is inertia?
7. Name the six simple machines. Give one example of each kind of machine.
8. What is the difference between conduction and convection?
9. When you take your clothes out of the dryer, they sometimes cling together. Why do you think this happens?
10. What are fossil fuels made out of? Is there an endless supply of fossil fuels? Name three other sources of energy we use to create electricity, power our factories, and keep ourselves warm.

Earth Science

Unit 4

Chapter 22
The Earth's Features

Chapter 23
The Earth's Crust

Chapter 24
The Earth's Atmosphere

Chapter 25
Weather and Climate

Chapter 26
The Earth's History

Chapter 27
The Earth's Oceans

Chapter 28
Astronomy and Space Exploration

Chapter 22

The Earth's Features

You might say the Earth is just another planet in our solar system. But it is actually a very special one. As far as we know, it is the only one with life on it. Why do you think there is life on Earth but not on the other planets?

Chapter Learning Objectives
- Name and describe the Earth's three layers.
- Explain how the Earth moves through the solar system.
- Explain time zones.

Words to Know

axis an imaginary line running from one pole, through the center of a planet, to the other pole

continents the seven major land masses on Earth: Africa, North America, South America, Asia, Europe, Antarctica, and Australia

core the center of the Earth

crust the thin, outer layer of the Earth

equator an imaginary line that circles the Earth halfway between the two poles; the 0 degree line of latitude

globe a map of the world that is round like a ball

lines of latitude lines drawn east and west on a map to help locate places

lines of longitude lines drawn north and south on a map to help locate places

mantle the layer of the Earth between the crust and the core

orbit a curved path

prime meridian the 0 degree line of longitude

solar system the sun and all the planets that revolve around it

Imagine that you have traveled through space to a distant part of the universe. You feel homesick for the Earth. So you try to find it by looking through a very powerful telescope. At first all you can see are millions of stars. But suddenly you see a small blue-green object. It is moving slowly on a path around one of the stars—the sun. As the little planet moves through space, it is spinning like a top. You are looking at the Earth!

In this chapter you will learn some facts about the planet Earth. You will learn what it is made of. You will also learn how it is positioned in the solar system.

Features of the Earth

The Earth is a planet in the solar system. The **solar system** is the sun and all the planets that circle it.

The surface area of Earth is 196,937,600 square miles. And the Earth has a mass of 5,882,000,000,000,000,000,000 tons.

Seventy percent of the Earth is covered with water. Water covers so much of our planet that it is sometimes called the "water planet." The main land masses on Earth are called **continents**. There are seven continents: Africa, Asia, Australia, Antarctica, North America, South America, and Europe. Oceans, mountains, plains, rivers, and islands are some of Earth's other physical features.

Solar means sun. The solar system is the system of planets revolving around the sun.

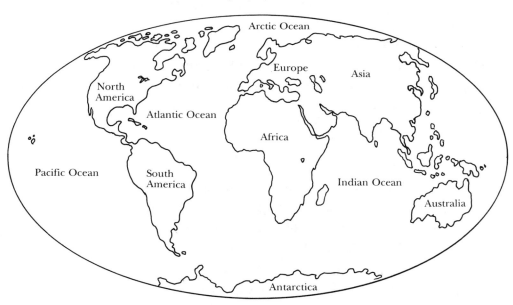

Can you find the United States of America on this map?

The northernmost part of the Earth is called the *North Pole*. The southernmost part is called the *South Pole*. An imaginary line circles the Earth halfway between the North and South Poles. This line is called the **equator**.

The highest point on Earth is the top of Mount Everest. This mountain rises to a height of 29,028 feet above sea level. Mount Everest is on the borders of two countries called Nepal and Tibet.

The lowest point of land on the Earth's surface is the shore of the Dead Sea. It is 1,310 feet below sea level. The Dead Sea is a very salty lake between the countries of Israel and Jordan. The Earth is the third planet from the sun.

The Age and Size of the Earth

The Earth was formed about four and one-half billion years ago. Scientists believe the Earth, along with the rest of the solar system, started off as a cloud of gases and dust. Over time, these gases and dust became more and more dense. Very gradually, they formed the planets and the sun.

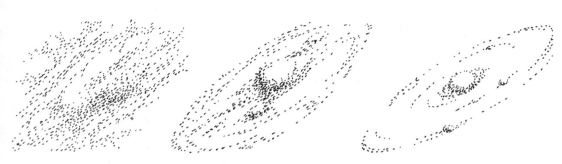

Most scientists agree that the solar system formed from gas and dust. This process has taken about five billion years.

The Earth is not a perfect sphere. A sphere is the shape of a ball. But the Earth bulges at the equator. The diameter of the Earth at the equator is 7,926 miles. The diameter of the Earth from pole to pole is 7,899 miles.

The Three Layers of Earth

The Earth has three layers. The center of the Earth is called the **core**. The core is about 4,400 miles across. Because the inner part of the core is under great pressure, it is a very dense solid. It is probably made of nickel and iron. The outer part of the core is also thought to be made of iron and nickel. But it is liquid. The Earth's core is very, very hot.

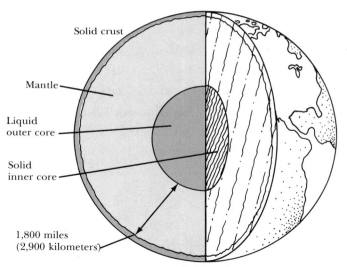

The earth has three layers: the core, the mantle, and the crust.

The middle layer in the Earth is called the **mantle**. The mantle is about 1,800 miles thick. Most of it is made up of solid rock. But the outer 60 miles of the mantle is a thick liquid. It flows very slowly. The mantle is made of silicon, oxygen, aluminum, iron, and magnesium.

The outer layer of the Earth, called the **crust**, is very thin. It is between 4 and 37 miles thick. The continents and the ocean floors are part of the Earth's crust. The crust is like a thin rocky shell around the mantle.

Movement of the Earth

The Earth moves in a curved path around the sun. This path is called an **orbit**. One full orbit around the sun is called a *revolution*. The Earth takes 365.25 days—one year—to *revolve* around the sun.

As the Earth orbits the sun, it also spins, or *rotates*, on its **axis** like a top. The Earth's axis is an imaginary line. It runs from one pole, through the center of the Earth, to the other pole. The Earth makes one complete *rotation* on its axis every 24 hours, or once a day.

The spin of Earth explains why there are days and nights. The part of Earth facing the sun gets the sun's light. When that part spins to the far side, the sun is blocked by the rest of Earth. So that side of the Earth would experience night.

The Earth spins on its axis from west to east. That explains why the sun appears to rise in the east and set in the west.

In what ways do you think life on Earth would be different if the Earth were not tilted on its axis?

Why Are There Seasons?

The four seasons result from the tilt of the Earth. The Earth is tilted 23-1/2 degrees on its axis. The tilt of the Earth causes two things to change during its orbit around the sun: 1) the number of hours of daylight; and 2) the angle at which the sun's rays strike Earth.

In the winter, the sun strikes the northern part of Earth for fewer hours than in the summer. Also, in the winter, the sun strikes the northern part of Earth at more of an angle than in the summer. This angle makes sunlight less strong than in the summer when the sun shines directly on the face of Earth.

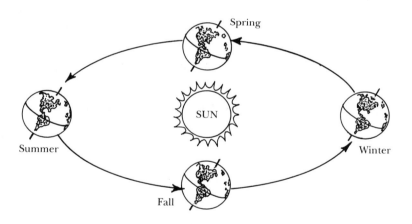

This diagram shows the earth's position at each of the four seasons. As the earth revolves around the sun, the tilt of its axis relative to the sun changes. This tilt determines the seasons.

Reading a Map of the World

A **globe** is a round world map. Globes and maps have lines running across them in both directions. These lines help people find the different features and places on Earth.

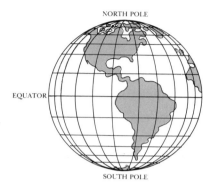

Latitude lines are horizontal. Longitude lines are vertical. The prime meridian is at 0 degrees longitude. The equator is at 0 degrees latitude.

The lines that run east and west are called **lines of latitude**. The equator is a line of latitude. The lines running north and south are called **lines of longitude**. These lines run from pole to pole.

Lines of latitude and longitude are assigned numbers called degrees. Lines of latitude are measured in degrees north and south of the equator. The equator is 0 degrees. It lies halfway between the poles. The North Pole is 90 degrees north latitude. The South Pole is 90 degrees south latitude.

The 0 degree line of longitude is called the **prime meridian**. It runs through Greenwich, England. The other lines of longitude are measured in degrees east and west of the prime meridian. Directly opposite the prime meridian on the globe is the 180 degree line of longitude.

Science Practice
Draw a picture of the world on a separate sheet of paper. Include the following features in your drawing. Label each of them.

North Pole South Pole Earth's axis
equator   continent   ocean

Amazing Science Fact
Mapmakers can now use computers to read photographs taken from airplanes and satellites. The results are very accurate maps. Also, the computers can add other kinds of data to maps, such as rainfall patterns or the location of dinosaur fossils. These new maps can be very helpful. For example, a city map that combines data on elevation, soil, and water tables can help city planners find the best sites for building.

Time Zones Around the World
The continental United States is divided into four time zones. They are called the Pacific, mountain, central, and eastern zones. Alaska and Hawaii are west of the Pacific zone and fall into different time zones. Each time zone is one hour different from the ones on either side. If you travel east, you *lose* one hour as you cross into each new time zone. Going west you *gain* one hour as you cross into each new time zone.

Suppose it is 7:00 A.M. in New York. New York is in the eastern time zone. At that same moment it would be 6:00 A.M. in Chicago; 5:00 A.M. in Salt Lake City; and 4:00 A.M. in Los Angeles.

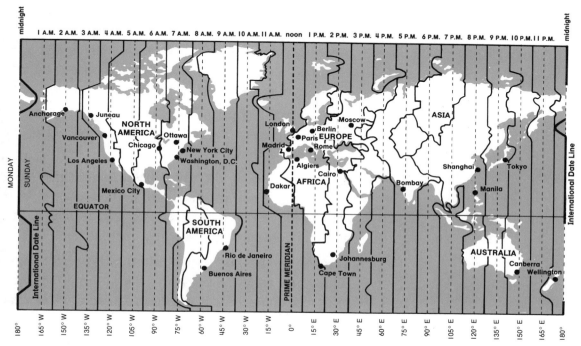

The time zones of the world

There are 24 time zones in the world. Each zone is about 15 degrees of longitude wide. The time zones are set up like this for a reason. The Earth rotates 15 degrees of longitude per hour.

The boundaries of the time zones are not always straight lines. They are drawn so that most states and small countries fit into one time zone.

Chapter Review

Chapter Summary

- The Earth is a planet in the solar system. About 70 percent of the Earth is covered with water. The Earth was formed about four and a half billion years ago. Scientists believe it was formed from a cloud of gases and dust.

- There are three layers in the Earth. The center layer is the core. The middle layer is the mantle. The thin outer layer is the crust.

- The Earth orbits the sun once a year. It spins on its own axis once a day. The seasons are caused by the tilt of the Earth.

- Lines of latitude and longitude are drawn onto maps. These lines help people locate certain features and places on Earth. The equator circles the Earth midway between the poles at 0 degrees latitude.

- People have divided the Earth into 24 time zones. Each time zone is one hour different from the ones next to it. Time zones help people keep track of time around the world.

Chapter Quiz

Answer the following questions on a separate sheet of paper.

1. Why is the Earth sometimes called the "water planet"?
2. Name the seven continents.
3. What are the three layers of the Earth? On which layer do you live?
4. How long does it take for the Earth to make one complete revolution around the sun?
5. How long does it take for the Earth to make one complete rotation on its axis?
6. What is another name for the 0 degree line of latitude?
7. What is another name for the 0 degree line of longitude?
8. Why are the boundaries of time zones not always straight lines?
9. Name the four time zones in the United States. Which time zone do you live in?
10. What causes seasons on the Earth?

Class Project: Studying the Globe

Use a globe to study the features of the Earth. Find as many of the following as you can.

oceans	continents	mountains
equator	prime meridian	rivers
islands	United States	your state

Mad Scientist Challenge: Time Zones

Answer these questions on a separate sheet of paper.

1. Suppose it is ten o'clock in the evening in New York. What time is it in California?
2. It is four in the afternoon in Oregon. Your friend leaves her office in Boston every day at five. Is it too late to telephone her there?

Chapter Twenty-Two

Chapter 23

The Earth's Crust

In 1906 a big earthquake hit San Francisco, California. Much of the city was destroyed. In 1989 the city was again struck by a strong earthquake. What can people in San Francisco do to get ready for the next earthquake?

Chapter Learning Objectives
- Describe the theory of plate tectonics.
- Explain what causes earthquakes and volcanoes.
- Explain how mountains are formed.
- Name the three main types of rock in the Earth's crust.
- Describe the processes of weathering and erosion.

Words to Know

erosion   the wearing away of soil by wind and water

geologist   a scientist who studies the history and structure of the Earth, especially as recorded in rocks

glacier   a large slow-moving field of ice

igneous rocks   rocks formed from magma

lava   what magma is called once it reaches the Earth's surface

magma   melted rock squeezed up from the Earth's mantle

metamorphic rocks   rocks that are formed when igneous or sedimentary rocks change under very high temperatures or pressure

plate   a large piece of the Earth's crust

plate tectonics   the theory that the Earth's crust is made of plates that slowly shift position

sedimentary rocks   rocks formed by many different rock particles "cementing" together

trench   a deep, long valley in the ocean floor

volcano   a hole in the Earth's surface through which magma pours from the mantle

weathering   the process that breaks down rocks and minerals

In California there are earthquakes every week. Most of them are so small that only scientists notice them. But every one hundred years or so a big one comes along and does a lot of damage.

Earthquakes are caused by the shifting of the Earth's crust. In this chapter you will learn how the Earth's crust moves. You will learn what causes volcanoes and earthquakes. And you will also learn what kinds of rock make up the crust.

Plate Tectonics

Most **geologists** believe that all the continents once formed one big continent called *Pangaea*. About 200 million years ago, pieces of land began breaking free from Pangaea. These land masses became the separate continents. Over the last 200 million years, the continents slowly drifted to their current positions.

Remember that a theory is a well-organized explanation of events or facts.

Plate tectonics is a theory that explains how parts of the Earth's crust move. This movement causes earthquakes, volcanoes, and the formation of mountains.

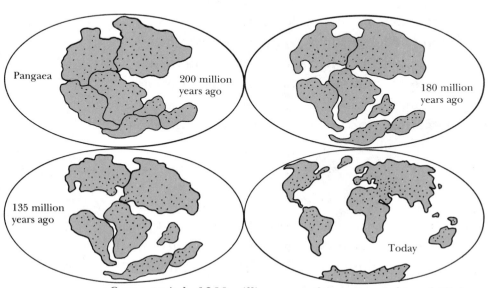

Over a period of 200 million years, the continents have drifted away from Pangaea to their current positions.

The theory of plate tectonics says that the Earth's surface is divided into **plates**. Six of these are very big, but there are several smaller plates, too. A plate is a piece of the Earth's crust. Plates can include both land masses and areas of ocean floor. At all times, these plates are moving very, very slowly.

The United States is on the American plate. It is drifting westward. Geologists believe they know why these pieces of the Earth's crust are moving. Remember that the outer part of the Earth's mantle is like a thick liquid of hot rock. It flows very slowly. Geologists say that the Earth's crust floats on this thick liquid. The plates are carried along in the flow of hot rock.

People in Science: Alfred Wegener

Imagine that you are a scientist looking at a globe of the Earth. Suddenly you notice that the continents look a lot like jigsaw puzzle pieces. They look as if they would all fit together. Maybe at one time they were part of one big continent, you think to yourself.

What would you have done with that idea?

A man named Alfred Wegener did have that thought in 1912. He wrote up his idea and presented it to other scientists. But most of them rejected his theory that the continents had once fit together as one. For one thing, he did not explain how the continents could move. But the recent theory of plate tectonics helps to support Wegener's theory.

Alfred Wegener

Colliding Plates Cause Trenches and Mountains

The plates of the Earth's crust meet at different places around the world. In some places, plates crash into each other. As they collide, one plate may get forced under the other. It gets pushed down into the mantle where the crust melts. When one plate gets pushed down under another, a **trench** forms. A trench is a deep, long valley in the ocean floor.

Sometimes two colliding plates pile up against each other. Then mountain ranges are formed. Mountains build up very, very slowly. Over millions of years, the plates push against each other. As they push, the land gets shoved upward, making mountains. The Alps in Europe were formed this way. So were the Sierra mountains in California.

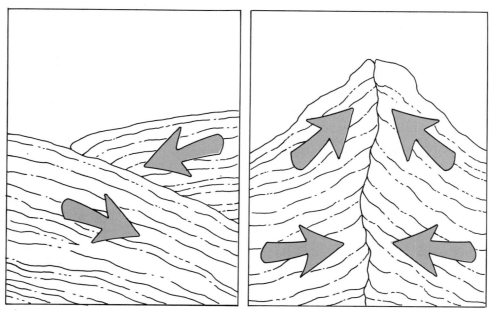

When two plates collide two things can happen. One plate can dive beneath the other, making a trench. Or the plates can pile up against each other, forming a mountain range.

Amazing Science Fact

Mount Everest, at 29,028 feet above sea level, is the highest point on Earth. But some mountains rise even higher than Everest from their base to their top. Mauna Kea, in Hawaii, is 33,474 feet high measured from the ocean floor next to it. But only 13,796 feet of Mauna Kea are actually above sea level.

Rubbing Plates Cause Earthquakes

Sometimes, instead of pushing into one another, two plates move against each other. This is true of the Pacific and American plates. They meet along the western coast of North America. The Pacific plate moves about two inches a year in a northwestern direction.

Imagine two gigantic pieces of the Earth's crust rubbing against one another. The plates do not slip by each other smoothly. Friction holds the upper layers of crust together. But the plates continue to move deeper down and pressure builds up on the surface. Finally, when the strain has more force than the friction can hold, the plates slip. The movement sends shock waves through the Earth. This is an earthquake.

Many of the earthquakes you see in the news occur along the west coast of the American continent. This is where the Pacific and American plates meet. Each time there is a sudden slip between these plates, an earthquake occurs.

Geologists are working on ways to predict earthquakes. So far, there are no sure ways to predict them, though.

When two plates move by each other earthquakes are the result.

Chapter Twenty-Three 263

Science Practice

On a separate sheet of paper, write true or false for each of the following sentences.

___ 1. Many geologists believe that all the continents were once attached.
___ 2. The theory of plate tectonics helps to explain earthquakes, volcanoes, and mountains.
___ 3. The United States is on the Eurasian plate.
___ 4. The Earth's plates float below the mantle.
___ 5. When two plates collide, they can form trenches or mountains.

Volcanoes

Sometimes, when plates move, openings are formed in the crust. Melted rock, called **magma**, gets squeezed up from the Earth's mantle. These openings are called **volcanoes**.

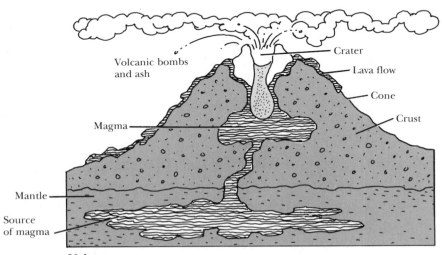

Volcano

Volcanoes can form on continents or on ocean floors. The magma, called **lava** when it reaches the surface, comes up through vents. Over time, the lava from a volcanic eruption builds up and creates a mountain. Volcanoes do not always erupt violently. Sometimes the magma flows quietly onto the surface.

When volcanoes occur on the ocean floor they can create islands. The Hawaiian Islands were formed this way.

There are a lot of earthquakes and volcanoes around the edge of the Pacific plate. This area is called the "ring of fire."

Amazing Science Fact

In the year 79 A.D. a volcano erupted near Pompeii, Italy. The volcano erupted suddenly and with great violence. It dumped tons of hot ash and cinders on the city of Pompeii. The ash hardened and preserved the shapes of many of the people who were killed by the volcano. Today, in Pompeii, you can see stone shapes cast by people who were surprised by the eruption.

Rocks and Minerals

The Earth's crust is made of rock. At some time in your life you may have collected rocks. If you did, you know that there are many, many different kinds. Some of the physical properties of rocks are color, shape, hardness, and texture. There are three basic types of rocks.

Igneous rocks are formed from magma. The magma is forced up from the mantle or lower crust. When it reaches the surface, the magma hardens into rock. Two kinds of igneous rock are *basalt* and *granite*. Igneous rocks make up about 95 percent of the Earth's crust.

Sedimentary rocks are made of many different rock particles cemented together. It takes a very long time for sedimentary rocks to form. Beds of clay, sand, or gravel may harden to make sedimentary rock. *Shale* is a kind of sedimentary rock made of hardened clay. *Sandstone* is another kind of sedimentary rock made of sand. *Coal* is a sedimentary rock formed from plant fossils. Sedimentary rocks are the most common rocks on the Earth's surface.

The third kind of rock is called metamorphic rock. **Metamorphic rock** is formed from sedimentary or igneous rock. When igneous or sedimentary rocks are made extremely hot, they turn into metamorphic rock.

All rocks are made of minerals. Minerals are non-living substances found in nature. There are at least 2,000 different kinds of minerals. Each one has a fixed chemical makeup. Many minerals are pure elements, but most are made up of combinations of elements. Some examples of minerals are gold, quartz, silver, and talc.

Weathering and Erosion

Weathering is the process that breaks down rocks and minerals. Water, ice, plants, animals, and chemical changes all play a part in weathering.

Think of a river. As the water moves along, it washes away little bits of rock from the riverbed. These bits of rock are swept downstream. As they move, they bump into other rocks and tumble along the bottom. Slowly, they break into smaller and smaller pieces.

Freezing also causes rocks to weather. First, water fills the cracks in rocks. Then when the water freezes, it expands. This can break the rock into smaller pieces.

Rain causes another kind of weathering. Raindrops beat on rocks gently but steadily like a million little hammers. Eventually, the rocks wear down.

Soil is an important product of weathering. As rocks break down into smaller and smaller pieces, the bits of rock mix with living things and their remains to form soil.

Rivers cause erosion as well as weathering. **Erosion** is the wearing away of soil. Very slowly, river valleys get carved deeper and deeper by the movement of water and rocks. Valleys cut by rivers are V-shaped.

Glaciers, or great fields of ice, cause another kind of erosion. They flow very slowly. As they move, they clear out everything in their paths. Valleys cut by glaciers are U-shaped.

Erosion is also caused by wind and rain.

Erosion and weathering are slow but powerful forces. Over millions of years, they can wear down entire mountain ranges.

Chapter Review

Chapter Summary

- Plate tectonics explains how parts of the earth's crust move. The theory of plate tectonics says that the Earth's crust is made up of several moving plates.

- Sometimes, when two plates collide, one plunges below the other, forming a trench. Other times, the two plates pile up to form mountains. When two plates move sideways against each other, the scraping causes earthquakes. Magma coming up through openings between plates causes volcanoes.

- The Earth's crust is mostly made of rock. The three main kinds of rock are igneous, sedimentary, and metamorphic. All rocks are made of minerals. Minerals are non-living substances with fixed chemical makeups.

- Weathering is the process that breaks down rocks and minerals. Water, ice, plants, animals, and chemical changes all play a part in weathering. Soil is an important product of weathering. Erosion is the wearing away of soil.

Chapter Quiz

Answer the following questions on a separate sheet of paper.
1. What is a "plate" in the theory of plate tectonics?
2. On what do the plates "float"?
3. What forms when two plates collide and one plate gets pushed down and under another?
4. What forms when two plates collide and pile up on each other?
5. What causes earthquakes?
6. What is magma called once it reaches the surface of a volcano?
7. Where do most earthquakes occur?
8. Name some islands formed by volcanoes.
9. What are the three main kinds of rocks on Earth?
10. Name at least five causes of weathering.

The Making of Valleys

On a separate sheet of paper, explain the difference between a U-shaped valley and a V-shaped valley. Describe how each type of valley is formed.

Mad Scientist Challenge: Fact and Theory

Do you remember the difference between scientific fact and scientific theory? A fact has been proven. A theory is an idea based on lots of good information. But a theory has not been proven beyond a shade of doubt. On a separate sheet of paper give one example of a scientific fact. Then give one example of a scientific theory.

Chapter 24

The Earth's Atmosphere

The cumulus clouds in this picture are a sign of fair weather. Yet they are made of water droplets. How do you suppose these water droplets remain suspended in air?

Chapter Learning Objectives
- Describe the makeup of the atmosphere.
- Explain how air pressure is related to winds and convection currents.
- Describe the different cloud forms and precipitation.

Words to Know

air pressure the weight of the gases pressing down on the Earth

atmosphere the gas that surrounds the Earth

barometer an instrument that measures air pressure

cirrus clouds high altitude clouds made of ice crystals

cumulus clouds large clouds that have flat bases and rounded masses piled on top

dew point the temperature at which water vapor turns to liquid water

humidity the amount of moisture in the air

ionosphere a layer in the upper atmosphere that begins in the mesosphere and extends upward through the thermosphere

mesosphere the third layer in the atmosphere

ozone a form of oxygen in a thin layer within the stratosphere

precipitation any moisture that falls from the atmosphere

stratosphere a layer in the atmosphere that begins about seven miles up; the second layer in the atmosphere

stratus clouds low foglike clouds that lie in a flat layer over a wide area

thermosphere the fifth layer in the atmosphere

troposphere the first layer in the atmosphere; the lower atmosphere in which most weather takes place

We tend to take the air around us for granted. Some people don't even think of the air as anything at all. In fact, air has mass. Air is matter. You can't see it, but it is made up of gases, dust, and water vapor. Sometimes the air is heavy and sometimes it is light.

In this chapter you will learn about some of the properties of air. And you will learn just how important air is to your life.

What Is the Atmosphere?

Take a deep breath. You just filled your lungs with some of the **atmosphere**. The atmosphere is the air that surrounds the Earth. The atmosphere is made up of a mixture of gases, fine dust, and water vapor. The pull of gravity holds the atmosphere to the Earth. The atmosphere travels with the Earth as the planet moves through space.

The atmosphere extends more than 1,400 miles above the Earth's surface. But it doesn't have a definite border. The atmosphere just gets thinner and thinner as it gets further from the Earth.

Seventy-eight percent of the atmosphere is nitrogen. Oxygen makes up about 21 percent of the atmosphere. There are also small amounts of argon, carbon dioxide, and other gases.

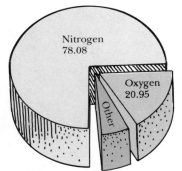

This graph shows the chemical makeup of the atmosphere.

Layers in the Atmosphere

Scientists divide the atmosphere into five layers. Some of these layers overlap. The layer nearest the Earth is called the **troposphere**. This is the layer in which we live. It is also the layer in which almost all weather happens. The troposphere is hottest near the Earth. It gets colder as it extends away from the Earth.

The next layer is called the **stratosphere**. Inside the stratosphere is a thinner layer called **ozone**. The ozone layer is made up of a form of oxygen. This layer is very important to living things. Ozone absorbs harmful radiation from the sun.

Science Alert

You may have heard about the ozone layer on the news. Scientists are very concerned about it. Certain kinds of air pollution may be breaking down this important oxygen covering. Without protection from the ozone, harmful radiation from the sun can get through to the Earth's surface. This radiation is an invisible form of light called *ultraviolet rays*. Some of this radiation passes through the ozone layer. It is the cause of sunburn. Scientists say an increase in this radiation could cause an increase in skin cancer in humans. It could have harmful effects on other life forms, too. We must find ways to cut back on air pollution in order to save the ozone layer—and possibly our lives.

The third atmospheric layer around the Earth is called the **mesosphere**. The fourth layer is called the **ionosphere**. The ionosphere begins in the mesosphere and goes upward through the fifth layer. The ionosphere contains many electrically charged particles. These electrically charged particles play a big role in radio communications. They reflect radio signals. The fifth layer is called the **thermosphere**. The air in the thermosphere is very thin.

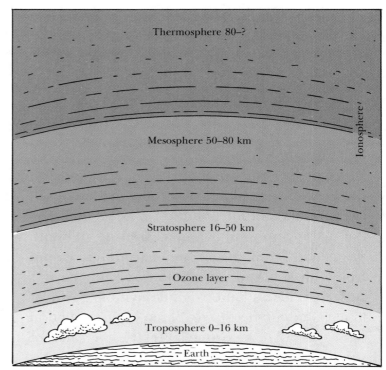

The five layers of the atmosphere.

Mountain climbers gasp for air as they near the top of a high mountain. That's because the atmosphere gets thinner the higher up they go. A climber needs to breathe in more air in order to get enough oxygen. Men and women who climb the highest peaks in the world must carry oxygen tanks.

What Is Air Pressure?

Gravity keeps the Earth's atmosphere from flying off into space. Gravity is strongest near the Earth's surface. For that reason, the parts of the atmosphere

Barometers are used by scientists to help predict the weather. A change in the barometric pressure often means a change in the weather.

closest to the Earth feel a stronger pull. This causes the air molecules to pack in more tightly. The result is that air close to the Earth's surface is denser. The farther away from Earth, the less dense the atmosphere.

Air pressure is the weight of the gases pressing down on the Earth. Air pressure changes all the time. Scientists measure air pressure with instruments called **barometers**.

Heating the Atmosphere

You have learned that energy from the sun reaches the Earth as radiation. As radiation strikes the planet, it changes to heat energy. Then the Earth radiates the heat back into the atmosphere. Water vapor and other gases in the air absorb the warmth.

Clouds block the sun. They keep much of the sun's energy from reaching the Earth. They also reflect sunlight back into the outer atmosphere. At night, however, clouds around the Earth act as a blanket. They trap the heat and keep it from escaping into outer space.

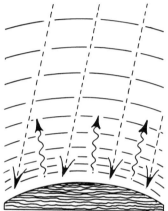

Energy from the sun warms the surface of the Earth and the Earth's surface warms the atmosphere.

What Causes Wind?

Remember that heat causes the molecules in matter to move faster and spread out. In other words, heat causes matter to become less dense. When air becomes less dense, it rises. Colder air sweeps into the empty space left by the warmer air. This sets up a *convection current*. Most winds are caused by convection currents.

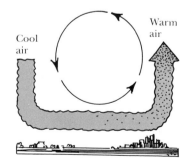

Convection currents result from the uneven heating of the atmosphere. Warm air rises. Cool air takes its place.

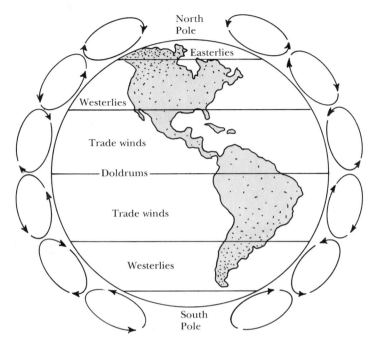

World wind systems

Warm air at the equator rises and moves out toward the poles. Cooler air flows toward the equator from both the north and south. This cooler air replaces the rising warm air, causing a huge convection current. But the Earth's own rotation breaks up this big wind system into several smaller, circular wind systems.

Winds are always named for the direction from which they come. So a *westerly* is blowing from the west, toward the east. A *north wind* is blowing from the north, toward the south. The picture of the Earth on this page, shows trade winds, westerlies, and polar easterlies.

Science Practice

Answer the following questions on a separate sheet of paper.
1. About how far does the atmosphere extend above the surface of the Earth?
2. What substance makes up most of the atmosphere?
3. Which layer of the atmosphere plays an important role in the transmission of radio waves?
4. What instrument is used to measure air pressure?

Local Winds

The Earth has several major wind systems. But local features often create smaller wind systems. Mountains, valleys, and big bodies of water all affect wind.

Cold mountain breezes blow down mountain slopes at night. Warm valley breezes blow up the mountain slopes on sunny days.

Like all winds, sea breezes are caused by uneven heating of air by the sun. Land heats up faster than oceans or large lakes. So air over land becomes warmer than air over bodies of water. When the warm land-air rises, the cold ocean or lake-air rushes in. This local wind is called a *sea breeze*.

At night the opposite happens. The land cools off faster than the ocean. So air over land will be cooler than air over ocean. The ocean air rises and the land air rushes out. This creates what is called a *land breeze*.

Sea breeze

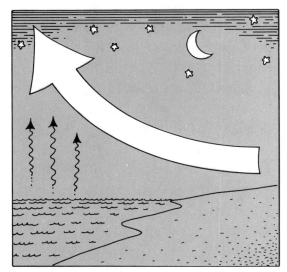
Land breeze

Water in the Atmosphere

There is water in the atmosphere. When the water is a gas, it is called water vapor. Vapor is composed of tiny droplets of water floating in the air as fog or mist. In its liquid form, water is rain. In its solid form it is sleet, snow, or hail.

Water gets into air by evaporating off soil, lakes, oceans, rivers, plants, and animals. The amount of water in the air is the level of **humidity**.

Water vapor in the air can change into liquid water if it is cooled enough. The temperature at which water vapor turns to liquid water is called the **dew point**. If air is cooled below the dew point, some of the water vapor forms tiny droplets of water. These droplets collect to make clouds. Fog is a low-lying layer of cloud.

When droplets become large and heavy enough, they fall to the Earth as rain. If the temperature in a cloud is below freezing, the water vapor may crystalize and fall as snow. **Precipitation** is any moisture that falls from the atmosphere.

Hail is another kind of precipitation. It is made up of hard pieces of ice. These form as water freezes around ice crystals that are moving through rain clouds. Sleet is rain that freezes as it falls through a layer of cold air near the ground.

Cirrus clouds

Clouds

Clouds form when water droplets or ice particles collect in the atmosphere. High humidity and cold temperatures help clouds to form.

There are three main kinds of clouds. **Cirrus clouds** form at high altitudes. They are made of ice crystals. Cirrus means "curled." These clouds are thin and feathery. They are usually bright white. Cirrus clouds are often seen in the mountains.

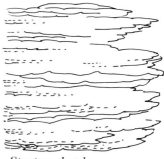
Stratus clouds

Stratus clouds are made of water droplets. Stratus means "spread out." These clouds are broad, flat layers of gray-colored clouds. Stratus clouds float low in the sky. They usually appear as an unbroken cloud cover. Stratus clouds are often a sign of stormy weather.

Cumulus clouds also form at low altitudes and are made of water droplets. Cumulus means "heaped up." These clouds are big, puffy masses. They look like cotton. Cumulus clouds are usually a sign of fair weather. But a thunderhead is a special kind of cumulus cloud. Thunderheads often bring rain.

Cumulus clouds

Chapter Review

Chapter Summary

- The atmosphere is a mixture of gases, fine dust, and water vapor that surround the Earth. Air pressure is the weight of these gases pressing down on the Earth.

- Warm air near the equator rises. Cold air from the poles sweeps in to fill the space. This sets up a convection current. Global winds are caused by convection currents. Local winds are caused by features of the land such as mountains, valleys, and large bodies of water.

- Humidity is the amount of moisture in the air. Precipitation is any kind of moisture that falls to the Earth. Rain, snow, hail, and sleet are examples of precipitation.

- A cloud is a collection of water droplets or ice particles. High humidity and cold temperatures help to form clouds. The three main kinds of clouds are cirrus, stratus, and cumulus. Fog is clouds lying near the ground.

Chapter Quiz

Answer the following questions on a separate sheet of paper.
1. What is the atmosphere made of?
2. What force keeps the atmosphere from escaping into space?
3. Which layer of the atmosphere is closest to the Earth?
4. Why is the ozone layer so important?
5. Why are the parts of the atmosphere closest to the Earth denser than the upper atmosphere?
6. How do clouds help keep the Earth warm at night?
7. Why does air near the equator rise?
8. Name four kinds of precipitation.
9. What are clouds made of?
10. What two weather factors help form clouds?

Charting the Clouds

Copy this chart on a separate sheet of paper. Write in as many properties for each kind of cloud as you can.

Clouds	Properties
Cirrus	
Stratus	
Cumulus	

Chapter 25

Weather and Climate

Winds within a tornado may be swirling as fast as 300 miles per hour. These winds may be moving upward as fast as 200 miles per hour. What do you think happens on the ground when a tornado touches down?

Chapter Learning Objectives
- Explain the difference between weather and climate.
- Describe how air pressure affects the weather.
- Describe several kinds of storms.

Words to Know

climate the average weather in a region over many years

cumulonimbus a tall, thick, white cloud with a dark base, often known as a thunderhead

cyclone an area of low air pressure with circular wind motion

front the place where two air masses of different temperatures meet

hurricane a stormy tropical cyclone over the North Atlantic Ocean

meteorology the scientific study of the Earth's atmosphere

occluded front the front that forms when a cold air mass overtakes a warm air mass

tornado a cyclone with violent whirling winds forming a funnel-shaped cloud that extends downward from a cumulonimbus cloud

typhoon a stormy tropical cyclone over the Pacific Ocean

Some people live where palm trees grow. They enjoy warm weather all year long. Others live where it snows throughout most of the year. Life is very different for people in these different climates.

Weather and climate play a big part in our lives every day. In this chapter you will learn the difference between weather and climate. You will learn about different kinds of weather. And you'll also learn how scientists predict weather.

What Is Weather?

Weather is the condition of the atmosphere at a certain time and place. Temperature, air pressure, humidity, wind speed, clouds, and precipitation are all parts of weather.

An *air mass* is a huge body of air covering a land or ocean area. Air masses can have high or low humidity. They can also be cold or hot. And as air masses move, they can change temperature or humidity.

Air masses are affected by the temperatures of the land or water below them. An air mass from the North Pole will be very cold. An air mass from a tropical area will be very warm.

Air masses do not mix. When two air masses come together, the place where they meet is called a **front**. Clouds and precipitation often form at fronts.

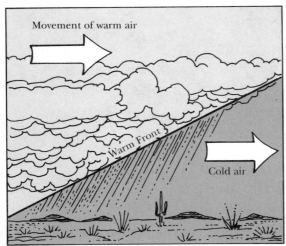

A warm front forms when a warm air mass moves into a cold air mass.

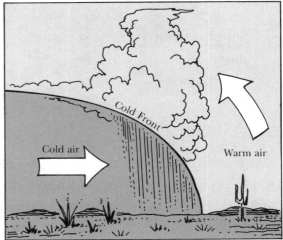

A cold front forms when a cold air mass moves into a warm air mass.

When a warm air mass moves into a cold air mass a *warm front* forms. As the warm air rises over the cold air it cools and releases moisture. There is usually rain or snow in a warm front. After a warm front passes, temperatures often rise.

When a cold air mass moves against a warm air mass a *cold front* forms. The cold air drives forward, forcing the warm air up. Then there are short but heavy showers of rain or snow. Temperatures drop after the cold front passes.

Cold fronts move much faster than warm fronts. So sometimes a cold front overtakes a warm front. This is called an **occluded front**. An occluded front produces less extreme weather than either a cold front or warm front.

Fronts that form over water carry more moisture than those that form over land. Ocean fronts are more likely to bring precipitation.

Thunderstorms

Thunderstorms are storms with rain, thunder, and lightning. They usually do not last very long. They are caused by warm, moist air rising quickly. The warm air may be forced upward by a mountain or a cold front. The warm and humid air cools quickly as it rises. As the moisture in the air cools it turns into rain.

In some places thunderstorms also occur on hot summer afternoons. This happens when moist air gets heated by the Earth's surface. As the warm air rises and then cools, it forms cumulus clouds. More warm air, called an updraft, blows up through the clouds. The updraft changes the cumulus cloud into a **cumulonimbus** cloud. Cumulonimbus clouds are tall and thick white clouds with dark bases. You may know them as *thunderheads*.

Thunderstorm

Electric charges build up in thunderheads. When these electric charges discharge, they cause lightning. The heat from lightning suddenly expands the air. This causes the loud noise called thunder.

Highs and Lows

If you watch the weather report on TV, you have heard about *highs* and *lows*. A high is an area where the air pressure is very high. A low is an area where the air pressure is very low.

Air moves from high pressure areas to low pressure areas. Moving air is wind. Usually winds blow out from the center of a high and into a low. Most of the time, high air pressure means good weather. Low air pressure means bad weather. Almost all storms are caused by lows.

Air moves from high pressure areas into low pressure areas.

Low Pressure Storms

A **cyclone** is an area of low air pressure. The low pressure is caused by warm air rising. Winds blow toward the low pressure area to fill it in. As the winds blow, they spin around the center of the low. Cyclones travel as they spin. They move across land at 500 to 1,000 miles a day.

A **hurricane** is a stormy tropical cyclone that forms over the North Atlantic Ocean. Hurricanes get their energy from warm tropical ocean water. As hurricanes move into colder northern waters, or over land, they usually weaken. But they can do a lot of damage along the coast. And they often cause flooding inland. A storm has to have winds of at least 75 miles per hour in order to be called a hurricane. But hurricane winds of more than 130 miles per hour have often been recorded. Tropical cyclones that form over the Pacific Ocean are called **typhoons**.

Science Alert

The very center of a hurricane is called the *eye*. The eye is usually about 20 miles across. There it is very calm. The sun might even shine in the eye of a storm. The air pressure is extremely low there. But don't be tricked! Hurricanes move very fast. If you are standing in the eye of a hurricane, you could be swept off your feet by racing winds.

A **tornado** is a cyclone that extends down from a cumulonimbus cloud. Tornadoes look like very dark funnels. The spinning winds sometimes reach speeds of 300 miles an hour. The updraft within the tornado may move as fast as 200 miles an hour.

Tornadoes do not cover as much ground as ordinary cyclones. But tornadoes are much more violent. Most of the tornadoes on Earth occur in the United States. The Great Plains and Mississippi Valley have many tornadoes each year.

In some parts of the United States tornadoes are a real danger. They can turn over cars and tear up houses.

Chapter Twenty-Five 287

Amazing Science Facts

Fog at airports can be very dangerous and inconvenient. Pilots can't see. Flights are canceled. People miss their meetings and connections.

Scientists have found some ways to get rid of fog. One way is to add heat to the atmosphere and stir it with big wind machines. Sometimes this breaks up fog.

Another way to break up fog is to "seed the clouds." Scientists put crystals of silver iodide in the clouds. This causes the moisture in the clouds to collect into drops and fall. Cloud seeding is also useful during dry spells. It causes clouds to produce more rain than they would otherwise.

Hurricanes get their energy from low pressure zones over warm tropical waters. Suppose these warm waters were chilled while the hurricane was still small. A lot of the hurricane's energy would be lost and it might do a lot less damage. Some scientists suggest towing icebergs into the paths of hurricanes. Others suggest pumping cold water from the bottom of the sea to the surface.

Science Practice

Number a separate sheet of paper from 1 to 5. Match each word with its meaning. Write a letter next to each number.

1. ___ typhoon
2. ___ cyclone
3. ___ front
4. ___ hurricane
5. ___ tornado

a. the place where two air masses meet
b. a cyclone that extends down from a cumulonimbus cloud
c. a tropical cyclone over the Pacific Ocean
d. an area of low air pressure
e. a tropical cyclone that forms over the North Atlantic Ocean

Weather Forecasting

Weather forecasters have a hard job. A lot of people depend on them. They can be wrong sometimes but they are more often right. Weather forecasting is a science.

The scientific study of the Earth's atmosphere is called **meteorology.** Meteorologists measure air pressure, humidity, wind speed and direction, air temperature, and the amount of precipitation.

There are meteorologists in weather stations all over the country. Several times a day they gather weather information. They send their information to the United States Weather Service. There the information is plotted on huge maps. These maps are updated several times a day. Other meteorologists study the maps. Sometimes they use computers to help them analyze their data. Then they make weather forecasts.

Meteorologists use weather stations like this one all around the country.

Try This Experiment

Study the weather for five days. On a separate sheet of paper, make a chart like the one below. Take notes each day. Then, on Friday, forecast the weekend weather. On Monday discuss whether you were right or not.

	M	T	W	Th	F
temperature					
precipitation					
clouds, if any					
other notes					

What Is Climate?

Weather is the day to day condition of the air. **Climate** is a description of the typical or average weather in a region over many years.

Many factors contribute to an area's climate. The latitude of the region is very important. Regions near the equator get the most direct sunlight. The rays of the sun strike the Earth head-on at the Equator. Near the poles, the rays spread out over a much larger area. So they are much less intense. Also, near the poles the sun's rays must pass through more atmosphere. The rays lose much of their energy.

Mountains also affect climate. They interrupt the flow of global winds. The land on the side of a mountain facing the wind gets a lot of rainfall. The wind and clouds get trapped on that side. The side of the mountain away from the wind is usually warmer and drier.

Large bodies of water also affect climate. Water moderates temperatures because it absorbs heat and holds it well. Places near water have milder temperatures throughout the year.

Ocean currents affect climate, too. The warm Gulf Stream current flows across the North Atlantic Ocean toward Europe. This current warms the winter air over northwestern Europe.

Three Main Types of Climate

The three main types of climate are tropical, polar, and temperate. *Tropical* climates are very warm and have no winter season. These climates often have heavy rainfall. Hawaii is an example of a place with a tropical climate.

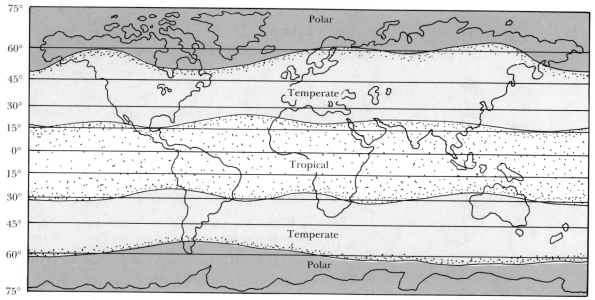

The three main types of climate

Polar climates have no true summer season. They have light precipitation, usually in the form of fine, dry snow. This is because the air is too cold to hold much water. Northern Alaska is a place that has a polar climate.

Temperate climates are warmer than polar climates. But they are cooler than tropical climates. Summers are warm and winters are cold. The precipitation varies. Most of the United States has a temperate climate.

People in Science: Benjamin Banneker

Benjamin Banneker was an African-American farmer and scientist who lived during the time of the American Revolution. Though most African-Americans were slaves at that time, Banneker was a free man and owned his own farm.

Banneker was excellent at math. He was so good that neighbors often came to him for help. His skill reached the ears of the new president, George Washington. Banneker was invited to help plan the new country's capital, Washington, D.C.

Banneker also taught himself astronomy. He corrected many mistakes that scientists of his day had made. Finally he decided he should make his own almanac.

In Banneker's day every family needed an almanac. This was a book of figures, information, and weather forecasts. Since most Americans were farmers, this book was very, very important.

Banneker farmed all day and studied all night. He used his compass and ruler to figure the movements of the sun and moon. He calculated exactly when and how high the tides would come in. He studied the stars with a telescope to calculate the rising and setting of stars and the position of the planets. He forecast the weather for the coming year based on his observations and past calculations. Banneker worked the complicated math problems over and over. He had to make sure every figure in his almanac was exact.

Banneker's almanac was a great success. Families all over the new United States used it to run their farms. By 1797, at least 28 different versions of Benjamin Banneker's almanacs had been printed in six years.

Chapter Review

Chapter Summary

- Weather is the day to day condition of the air. Climate is the average weather in a region over many years. The three main types of climate are tropical, polar, and temperate.

- Masses of air take on the characteristics of the land or water below them. When two air masses of different temperatures meet, a front forms.

- Thunderstorms are caused by warm, moist air rising quickly. Other storms are caused by low air pressure. Winds rush in to fill the low pressure area. A cyclone is a low pressure area. Hurricanes are stormy cyclones over the North Atlantic Ocean. Typhoons are stormy cyclones over the Pacific Ocean. Tornadoes are special kinds of cyclones that extend down from cumulonimbus clouds.

- The United States Weather Service plots weather information on huge maps several times a day. These maps are used for forecasting the weather.

Chapter Quiz

Answer the following questions on a separate sheet of paper.

1. Name six things that affect the weather.
2. What is an air mass?
3. What causes a warm front?
4. What is a cumulonimbus cloud and what kind of storm does it often accompany?
5. Are storms often caused by low air pressure or high air pressure?
6. How is climate different from weather?
7. People at the United States Weather Service receive information from weather stations all over the country. What do they do with this information?
8. What are the three main kinds of climate in the world? What kind of climate do you live in?
9. Name three physical features of the Earth that affect climate.
10. How do mountains affect the climate?

Key Weather Words

Number a separate sheet of paper from 1 to 3. Choose the correct answer to complete each sentence. Write a letter next to each number.

1. The weather is the condition of the atmosphere
    a. at the present time.
    b. over many years.
2. All weather fronts usually bring
    a. fair weather.
    b. cloudy and rainy weather.
3. An occluded front is formed when
    a. a warm front overtakes a cold front.
    b. a cold front overtakes a warm front.

Chapter 26

The Earth's History

The Grand Canyon is one of the most interesting and beautiful geological formations on Earth. It was formed by millions of years of weathering and erosion. How do you think it will look one million years from now?

Chapter Learning Objectives
- List the four geological eras.
- Explain two ways that geologists date rocks.

Words to Know

Cenozoic era the current geological era, which began 65,000,000 years ago

geological eras periods of time in the Earth's history

Mesozoic era the geological era beginning 225,000,000 years ago and lasting 160,000,000 years

Paleozoic era the geological era beginning 570,000,000 years ago and lasting 346,000,000 years

Precambrian era the geological era beginning 4,500,000,000 years ago and lasting nearly four billion years

radioactive dating finding the age of rocks by measuring the decay of radioactive elements

---

Sunrise on the Grand Canyon is a magnificent sight. The early sun strikes the many colors in the canyon walls. These walls are layered with many kinds of rock. Each different-colored layer is from a different period in history. You could say the walls of the Grand Canyon are striped with the history of the Earth.

In this chapter you will learn about the history of our planet. You will also learn how scientists date rocks and fossils.

The Age of Rocks

Layers of rocks are often arranged in order of age. The lowest layer is usually the oldest. The top layer is usually the youngest. But the Earth's shifting crust sometimes causes uplifting and folding. This can cause older layers of rock to rise above younger ones.

Radioactive dating is a more accurate way to determine the age of rocks. Some elements are radioactive. That means they give off nuclear radiation. As they do this, the element slowly breaks down into other elements. Each radioactive element decays at a certain rate. By looking at the degree of decay of certain elements, scientists can tell how long ago they formed. Radioactive elements act as geological clocks. The radioactive element uranium is often used for the radioactive dating of rocks.

Fossils and the History of Life

Geologists learn a lot from fossils. For example, fossils provide evidence that the Earth's continents and climates have changed a great deal. Fossils of tropical plants have been discovered close to the Arctic Circle. The remains of woolly mammoths, animals suited for the coldest climates, have been found in New York State.

Amazing Science Fact

Scientists have discovered shark teeth fossils on mountaintops. They believe that the fossils formed in rocks on the ocean floor. Over millions of years the Earth's plates moved. When some of them collided, new mountain ranges were formed out of what was once ocean floor. The shark teeth were thrust up with the new mountains.

Geological Time

The history of the Earth has been divided into four time periods. These time periods are called **geological eras**. They are the Precambrian era, Paleozoic era, Mesozoic era, and Cenozoic era.

Scientists study fossils to help them decide when each era began and ended. The rocks from each era have certain kinds of fossils in them. Each era marks big changes in landforms and climate in many parts of the world. At the end of each era many forms of life suddenly became extinct.

Scientists aren't sure what has caused mass extinctions. Many scientists believe the Earth's climate changed suddenly. This may have caused the extinctions.

Science Practice

Complete these sentences on a separate sheet of paper. Use the following words.

fossil   radioactive dating   geological eras
    uplifting and folding   mass extinction

1. _____ can change the order of rock layers in the Earth.
2. Each geological era has ended in a _____ .
3. A _____ is the remains of a plant or animal.
4. One way to tell the age of rocks is to use _____ .
5. The history of the Earth is divided into four _____ .

Precambrian Era

The **Precambrian era** started with the Earth's beginning about 4,500,000,000 years ago. The era lasted for nearly four billion years. This era covers about 85 percent of all geological time. There are still rocks from Precambrian time on every continent. In fact, some of the granite and marble used in building today was formed in the Precambrian era.

There are very few fossils from the Precambrian era. But scientists have found the remains of Precambrian algae and bacteria. They have also found fossils of tunnels made by worms. The worms probably made the tunnels in mud which hardened into rock.

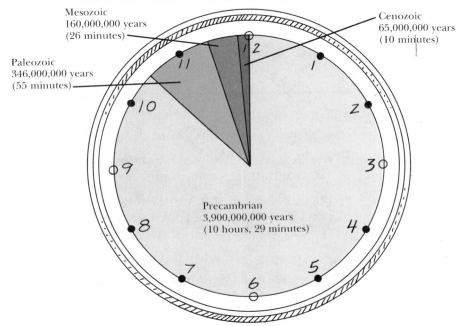

Suppose all the Earth's history were squeezed into twelve hours. This chart shows how much time each geological era would take up.

Paleozoic Era

The **Paleozoic era** began about 570,000,000 years ago. It lasted about 346,000,000 years. The Paleozoic era is known as the age of invertebrates and marine life. There are lots of fossils from this period. There are fossils of Paleozoic jellyfish, sponges, snails, seaweeds, and ferns.

Many coal beds have been dated to the Paleozoic era. Coal is made from the remains of plants. This suggests that the Paleozoic era was a very good time for plants. The Earth must have been warm and moist. There must have been many swamps in which giant plants grew.

Near the end of the Paleozoic era, many mountain ranges were formed. The Appalachian mountains in the eastern United States are an example. As parts of the crust rose up from sea level, the swamps dried up. Many of the ferns died out.

Mesozoic Era

The **Mesozoic era** is known as the age of reptiles. Dinosaurs and other reptiles were plentiful during this era. Flowering plants replaced the ferns that died at the end of the Paleozoic period. The Mesozoic era began about 225,000,000 years ago and lasted about 160,000,000 years.

There were many kinds of dinosaurs during the Mesozoic era. Some lived in the forests, while others lived in swamps. Still others lived on the open plains. While some dinosaurs were small and fast runners, others were huge and moved very slowly. Some dinosaurs ate meat and others ate only plants.

The planet was very dry at the beginning of the Mesozoic era. There were many volcanic eruptions. Later in the era, the Sierra Nevada mountains were pushed up. So too were the coast ranges of California.

Late in the Mesozoic era, there were fruit trees, willow trees, grasses and grains. Dinosaurs became extinct at the end of the Mesozoic era.

Plants flourished in the Paleozoic era.

One of the fiercest dinosaurs was more than thirty feet tall. But its brain was only as big as a chicken egg.

Dinosaurs flourished in the Mesozoic era.

Cenozoic Era

The **Cenozoic era** is the most recent era in geological history. It began about 65,000,000 years ago and runs up to the present. You are in it right now. This is known as the age of mammals and birds.

There have been several ice ages in the Cenozoic era. During each ice age, huge glaciers flowed south from the Arctic. The glaciers covered the northern parts of Europe, Asia, and North America. The ice was almost a mile thick in places. The last ice age ended about 12,000 years ago.

In 1849, gold was discovered in the Sierra Nevada mountains. People rushed to California from all over the country. They hoped to make their fortunes finding gold made in the Mesozoic era.

On the Cutting Edge

Scientists have drilled a deep hole in the ice of central Greenland. How deep? About 4,954 feet (1,510 meters). That is about halfway to the bottom of the Arctic ice sheet. They pulled a core of ice out of the hole. The ice core is 5.2 inches (13.2 cm) wide. The ice on the bottom of that core is 8,500 years old. So this ice core is a record of time back to the end of the last ice age.

The scientists are not done drilling yet. They hope to reach bedrock at the bottom of the ice. Then they will have a 200,000 year record. It will give them information on climate, atmosphere, and volcanic activity.

What Will the Next Era Be Like?

What is in store for Earth in the geological future? No one knows for sure, but scientists have many theories.

Some scientists think the Earth will experience another ice age. Others think that pollution will cause the atmosphere to heat up and melt the polar caps. One thing is for sure: things will not stay the way they are forever.

For one thing, the sun will eventually burn out. All stars do. Without the sun, there could be no life on Earth. Fortunately, there is no danger of the sun burning out in your lifetime or the lifetime of your children. In fact, it will probably be billions of years before the sun dies.

Chapter Review

Chapter Summary

- Geologists study layers of rocks in order to determine their age. Often, the top layers are younger than the underlying layers. But these layers can be mixed up by movements of the Earth's crust.

- Radioactive dating is a better way to determine the age of rock. Scientists measure the rate of decay in radioactive elements. This lets them know how old certain rocks are. Fossils also provide good clues to the Earth's history.

- Geologists have divided the history of the Earth into four main time periods. The Precambrian era makes up 85 percent of the planet's history. Only the simplest of life forms existed in this era. In the Paleozoic era invertebrates and marine life were plentiful. Many coal beds were formed at this time of rich plant life. The Mesozoic era is known as the age of reptiles. The Cenozoic era, the current one, is the age of mammals and birds.

Chapter Quiz

Answer the following questions on a separate sheet of paper.

1. What can geologists learn from fossils?
2. Why can't scientists be sure that the highest layers of rock in a canyon are the youngest?
3. Which of the four eras lasted the longest?
4. Which era is known as the era of invertebrates and marine life?
5. In which era did many coal beds form? What conditions on the Earth would have helped form these coal beds?
6. In which era were dinosaurs plentiful?
7. Which era are you in now? What kinds of animals dominate this era?
8. When was the last ice age?
9. What marks the end of each era?
10. What kind of fossils have scientists found from the Precambrian era?

Reporting on Science

Look up dinosaurs in an encyclopedia. Choose one kind of dinosaur to research. Find out where on the Earth that dinosaur lived. Did it eat meat or only plants? How big was it? Describe what it looked like.

Mad Scientist Challenge: Charting Geological Time

On a separate sheet of paper draw a chart of the four geological time periods. Write at least two things that describe each era.

Chapter 27

The Earth's Oceans

Many parts of the ocean floor have still not been explored. Some are so deep that sunlight does not reach them. Do you think anything can live in these deepest parts of the sea?

Chapter Learning Objectives
- Explain why the sea is important to life on Earth.
- Define and explain waves, tides, and currents.
- Describe three features of the ocean floor.

Words to Know

continental shelf   the gently sloping part of the ocean floor that begins at the edge of continents

continental slope   the steep slope of the ocean floor between the continental shelf and the ocean basin

mid-ocean ridge   an underwater mountain chain that runs down the middle of the ocean floor

ocean basin   the bottom of the ocean floor

oceanographer   a scientist who studies the oceans

salinity   a measure of saltiness

seismic sea waves   giant waves caused by earthquakes on the ocean floor

tides   the rising and falling of sea level, caused by the gravitational pull of the moon and the sun

undertow   a current of water that washes back out to sea under incoming waves

Except for a swim now and then, most people spend all of their time on land. People tend to think of land as the important part of Earth. But 70 percent of Earth is covered with ocean. And most of the Earth's organisms live in the sea.

The oceans are frontiers. A *frontier* is a place that has not been thoroughly explored. Only in the last century have scientists begun to look carefully into the deep and mysterious oceans. Scientists who study the oceans are called **oceanographers**. In this chapter, you will learn some of the things they have discovered.

What Is an Ocean?

Look at the world map on this page. There are five oceans: the Atlantic Ocean, the Pacific Ocean, the Indian Ocean, the Arctic Ocean, and the Antarctic Ocean. Notice how all the oceans are connected.

If you have ever tasted ocean water you know it is *salt water*. Ocean water also contains other chemicals such as chlorine, sodium, magnesium and calcium.

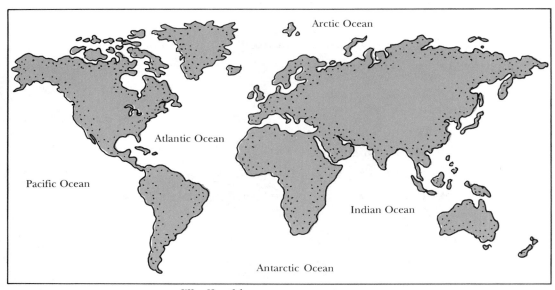

The Earth's oceans

Salinity is a measure of saltiness. Ocean water is usually between 3.3 and 3.7 percent salt. Salts are carried to the oceans by rainwater washing off the continents. As weathering and erosion break down rocks, salts in the rocks enter the soil. Then rainwater drains off the soil. As it does this, it dissolves *sodium chloride* (table salt) and other salts. These salts are carried to the sea by rivers.

Amazing Science Fact

The Dead Sea, the lake that borders Israel and Jordan, is 23 percent salt. Very little fresh water flows into the Dead Sea. Since the area is very warm, a lot of water evaporates off the surface. The salt remains. Very little plant life and no fish live in the Dead Sea. The salinity is too high.

Surface Ocean Currents

A river of warm water courses through the Atlantic Ocean. This river is thirty miles wide. It is a different color than the ocean water around it. And it is much warmer. This river of water is called the *Gulf Stream.*

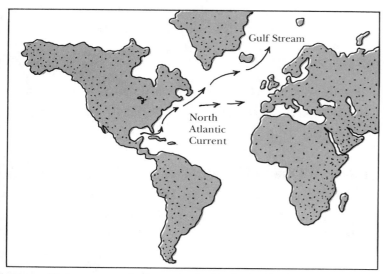

The gulf stream travels from the Gulf of Mexico to Newfoundland. There it meets the North Atlantic Current which moves east to England.

The Gulf Stream is warm because it flows from the Gulf of Mexico, which is near the equator. It sweeps north along the east coast of the United States to Newfoundland. Another current, called the *North Atlantic Current*, moves from Newfoundland across the Atlantic Ocean to England. Together the Gulf Stream and the North Atlantic Current make up the Gulf Stream system.

Ships crossing the Atlantic toward Europe try to stay in the Gulf Stream system. It helps them move faster. On their way back to the United States, ships avoid the Gulf Stream system. They don't want to have to sail against the Gulf Stream's current.

The Gulf Stream and the North Atlantic Current are ocean currents. An *ocean current* is a mass of water that flows like a river through an ocean. Currents are caused by areas of wind that blow steadily in the same direction. These winds pull water along with them.

Winds start currents. But the rotation of the Earth also affects the direction of currents. It causes them to move in circular patterns.

Undersea Ocean Currents

Wind-driven currents are all *surface currents*. That is, they flow near the surface of the oceans. Another kind of current is the undersea current. These flow below the surface currents. Undersea currents are caused by density differences in ocean water.

Cold water is more dense than warm water. Salt water is more dense than fresh water. Because it is more dense, cold salty water will sink. Warm, less salty water will rise. This process starts a convection current.

One convection current occurs between the poles and the equator. Water near the poles is much saltier than water near the equator. It is also much colder. This dense water sinks as it moves toward the equator. It becomes an undersea current below the warm, less salty water over the equator. The warm water from the equator flows toward the poles.

Every current, whether surface or undersea, has a counter current. For example, the Gulf Stream has an undersea current flowing beneath it. And the undersea currents flowing from the poles have warm surface currents flowing above them. The surface and undersea currents always flow in opposite directions.

Try This Experiment

Get some *cold* saltwater. Also get some *warm* plain water. Add a couple drops of food coloring to the plain water. Gently pour both kinds of water into a clear bowl. Explain what happens.

Science Practice

Number a separate sheet of paper from 1 to 6. Decide whether each statement is true or false. Write *T* or *F* next to each number.

1. ___ Oceans cover about 25% of the Earth.
2. ___ The Dead Sea has high salinity.
3. ___ The Gulf Stream is a warm, surface current.
4. ___ All currents are caused by earthquakes.
5. ___ Warm water sinks and cold water rises.
6. ___ Undersea currents are caused by convection currents.

Ocean Waves

Most ocean waves are caused by local winds. But underwater earthquakes and volcanoes sometimes cause waves, too.

Off the shore, the water moves up and down as the waves move forward. If you watch a bird sitting on the sea, you will see that it rises and falls as waves pass. It is not carried forward by the wave. But the energy in the wave moves forward.

Water in waves does not move forward. But the energy in waves does move forward.

Have you ever seen people surfing? What kind of waves do you think surfers like best? What conditions at the shore might make for good surfing waves?

Near the shore, waves "drag" on the ocean floor. This causes the tops of waves to spill over. The waves at the shore are called *breakers*.

After a wave breaks, it slides back to the ocean under the breakers. This backward movement of ocean water at the shore is called **undertow**.

Seismic Sea Waves

Once in a while, a giant wave crashes to shore. These giant waves are caused by earthquakes in the sea floor. Many people call these waves "tidal waves." But they don't have anything to do with the tides. Their scientific name is **seismic sea waves**. The word "seismic" means "having to do with earthquakes."

Ships at sea often don't even notice seismic sea waves passing. In deep water the waves are only about one or two yards high. But as they reach shore, these waves can be over a hundred feet high. Seismic sea waves can travel faster than 350 miles per hour. They can do a lot of damage when they crash into the shoreline.

Tides

Ocean waters rise and fall twice a day. These up and down movements are called tides. *High tide* is the point at which the water is as high as it gets. *Low tide* is the point at which the water is as low as it gets.

Tides are caused by the gravitational pull of the moon and sun. Gravity actually pulls the water back and forth as the Earth spins on its axis. The sun is much farther away from the Earth than the moon. So the sun's gravitational pull on Earth is not as strong as that of the moon.

Twice a month the sun and the moon lie in a straight line with the Earth. This makes the gravitational pull especially strong. So the tides are especially high and low. These tides are called *spring tides*.

Twice a month, the sun and moon form a right angle with the Earth. At these times, the sun's pull is working against the moon's pull. That means the tides are neither very high nor very low. They are called *neap tides.*

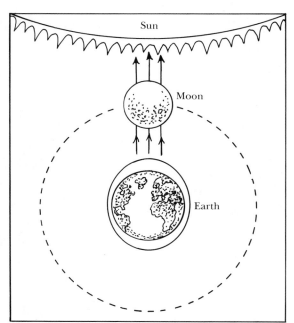

Spring tides: the sun and the moon pull together to make extra high and low tides.

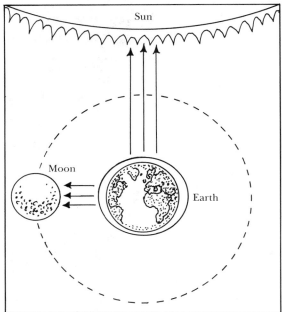

Neap tides: gravity from the moon and the sun work against each other. High and low tides are less extreme.

The Floor of the Sea

On the floor of the sea there are plains, mountain ranges, volcanoes, trenches, and plateaus.

Huge mountain ranges run down the middle of the major oceans. These are called **mid-ocean ridges**. Scientists believe that new crust is made in these underwater mountain ranges. They say that hot magma pushes up through openings in the ridges.

The magma then cools and hardens into rock on either side of the ridge. This causes the slow but steady movement of the sea floor crust away from the ridge. This process is called *sea floor spreading*. Most underwater earthquakes occur along the mid-ocean ridge.

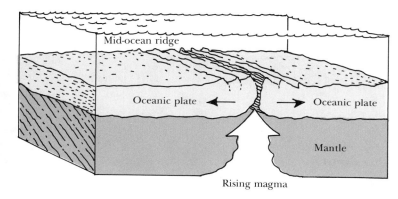

As plates move apart, magma fills the gap between them forming a mid-ocean ridge.

If the sea floor is spreading, are the oceans getting bigger and bigger? No. Remember what you learned in Chapter 23 about plate tectonics. Sometimes one plate slips under another plate. The lower plate melts and returns to the magma under the Earth's crust. This happens in the oceans as well as on land. The places where one plate slips under another are called trenches. They are the deepest parts of the ocean.

The gentle slope from the shore out to sea is called the **continental shelf**. The slope gets steep before evening out. This steep part is called the **continental slope.** The continental slope leads to the **ocean basin**, the bottom of the ocean floor. The mid-ocean ridge and trenches are on the ocean basin.

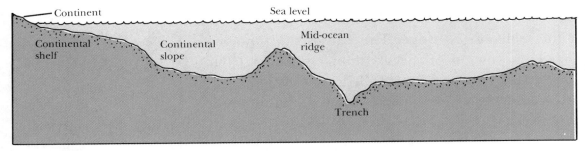

The sea floor

Sound waves are used to measure ocean depth. Oceanographers send a sound wave to the sea floor. When the sound wave hits bottom it bounces back. The oceanographers measure the amount of time it takes for the sound wave to bounce back to the ship. The machine they use is called a sonic depth recorder.

Ocean Resources

The sea is full of riches. For thousands of years people have fished the sea for food. Seaweed (a kind of algae) is harvested for many different products. It is used in cosmetics and to thicken milk shakes. It is also used in some paints, dyes, and papers.

People are experimenting with getting energy from the tides. The ocean floor is also rich in oil, natural gas, and some important metals. Salt and chlorine are both taken from the sea. In some places, people even remove the salt from ocean water to make drinking water.

Ocean algae make up to 90 percent of the food produced by living organisms on Earth. They are the first step in many, many food chains. People would not have food if it weren't for algae. And algae produce something else essential for life: oxygen. Up to 90 percent of all the oxygen in the air is produced by the photosynthesis of ocean organisms!

People in Science: Sylvia Earle

Sylvia Earle is a marine botanist and deep-sea explorer. She has spent 6,000 hours of her life—over seven months—under water. In 1970, she was the leader of an all-women team that lived in a tank underwater for two weeks. They spent hours swimming outside the tank and studying the ocean wildlife. Earle discovered 26 plants that no one had seen before in that part of the ocean.

Earle is also known for having made the deepest dive ever. She went down to 1,250 feet. This is very dangerous. The pressure of the water at that depth is over 600 pounds per square inch. If she'd gotten a leak in her plastic and metal suit, the water would have crushed her to death.

Sylvia Earle

Earle loves everything about the ocean. She says, "People are under the impression that the planet is fully explored, that we've been to all the forests and climbed all the mountains. But in fact many of the forests have yet to be seen for the first time. They just happen to be underwater."

Earle has a strong interest in all living things. She shares her home with two dogs, six cats, a parrot, a macaw, two geese, an iguana, several snakes, some fish, some tarantulas, a caiman, and an alligator.

Chapter Review

Chapter Summary

- All the oceans on Earth are connected. Ocean water is salt water. The main difference between ocean water and lake and river water is the amount of salt in it.

- An ocean current is a mass of water that flows through the ocean like a river. Surface currents are driven by wind. Undersea currents are caused by the sinking of dense cold water and the rising of warm water. All currents have a counter current, either below or above them.

- Most ocean waves are caused by local winds. Water in waves does not move forward. But the energy in waves does move forward. At the shore, waves drag on the ocean floor. This causes them to break.

- Tides are the periodic rise and fall of ocean water. They are caused by the gravitational pull of the sun and moon.

- The floor of the ocean is varied. There is one major mountain range running down the middle of all the oceans. It is called the mid-ocean ridge. There are volcanoes, trenches, and plateaus on the ocean floor as well.

- The sea is rich in resources. Life forms in the ocean are the start of many food chains. They are also responsible for creating most of the oxygen in the Earth's atmosphere.

Chapter Quiz

Write answers to these questions on a separate sheet of paper.

1. Name the four biggest oceans on Earth.
2. Where does the salt in the sea come from?
3. What is an ocean current?
4. Why do ships try to travel from the United States to Europe in the Gulf Stream system?
5. Why do you think certain animals like to live in the Gulf Stream?
6. What causes surface currents?
7. What causes undersea currents?
8. What causes seismic sea waves?
9. Where is new crust for the Earth being made?
10. Name five resources produced by the sea.

Exploring the Sea Floor

Complete these sentences on a separate sheet of paper.

1. The deepest places in the ocean are called _____ .
2. The depth of the ocean is measured by using _____ waves.
3. The mountain range in the oceans is called _____ .
4. The first part of the sea floor going out from the continent is called the continental _____ .

Mad Scientist Challenge: Food from the Sea

Go to a large grocery store. Take a notebook and a pencil. Find as many foods from the sea as you can. Write their names. In addition to fish, try to find some seaweed or algae products.

Chapter 28

Astronomy and Space Exploration

Saturn is the solar system's second largest planet. Saturn's system of rings makes it one of the most beautiful planets. Do you know what the rings around Saturn are made of?

Chapter Learning Objectives
- List the nine planets in the solar system.
- Describe at least two important events in space exploration.
- Explain the difference between meteors and meteorites.

Words to Know

asteroids   rocky fragments that circle the sun between the orbits of Jupiter and Mars

astronomers   scientists who study the universe beyond the Earth's atmosphere

comet   a ball of ice, dust and gases that orbits the sun

meteor   a rocky fragment from space that burns up when it enters the Earth's atmosphere

meteorite   a rocky fragment from space that lands on the Earth

satellite   an object that revolves around a planet. Both moons and man-made objects put into orbit are satellites.

People have walked on the moon. They have sent spacecraft to several other planets. **Astronomers** have studied many of the objects found in space. But they will never see all of it. Outer space is so vast, it will always remain a frontier.

In this chapter you will learn some of the things that astronomers have discovered. And you will learn where the Earth fits into the big picture of the universe.

The Universe

The universe is all the matter, space, and energy that exists. The Earth is only a tiny speck in the vast universe.

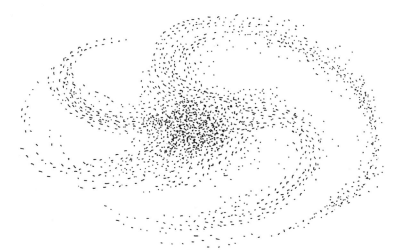

The sun is one of billions of stars in the Milky Way galaxy.

A *galaxy* is a group of stars, dust, and gases held together by gravitational force. In Chapter 1 you learned that the Earth is part of the galaxy called the *Milky Way*. This galaxy is shaped like a pinwheel. It has spiral arms that swirl around a center. The sun is in one of these arms. There are more than 100 billion stars in the Milky Way galaxy. And there may be more than a billion galaxies in the universe.

The Solar System

The sun is a star. All stars are big balls of very hot burning gases. They give off their own light. The core of the sun is made up of hydrogen and helium gases.

The *solar system* is made up of the sun and all the planets and other objects that revolve around the sun. Each of these objects moves around the sun in its own path. This path is called an *orbit*.

You know that the sun is just a medium-sized star. Why do you suppose it looks so much bigger and brighter than other stars?

Amazing Science Fact

Sunspots are dark areas on the surface of the sun. They are caused by areas of gas that are cooler than most of the sun. Sunspots appear to move across the sun the way storms move across the Earth. They can last from a few hours to a few months. Some of these spots are much larger than the Earth.

The sun is much bigger than anything else in the solar system. The sun's mass is 700 times greater than that of all the planets put together. Because it is so big, the sun has a very strong gravitational pull. The sun's gravitational force keeps the planets and other objects in orbit around it.

Planets in the Solar System

Other than the sun, the planets are the biggest bodies in the solar system. The planets do not orbit the sun in perfect circles. Their orbits are *elliptical*. This means that sometimes a planet is closer to the sun than at other times.

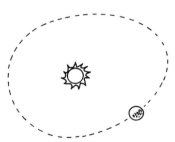

Most orbits are not perfectly circular. Most orbits are elliptical, or egg-shaped.

The Earth orbits the sun every 365 and 1/4 days. It takes Pluto 248 years to circle the sun once. It only takes Mercury 88 days. Look at the picture of the solar system on page 324. Why do you suppose it takes Pluto so much longer than Mercury to orbit the sun?

An *axis* is an imaginary line running through a planet from one pole to the other. Every planet spins on its own axis. A planet's "day" is the time it takes to make one complete turn on its axis.

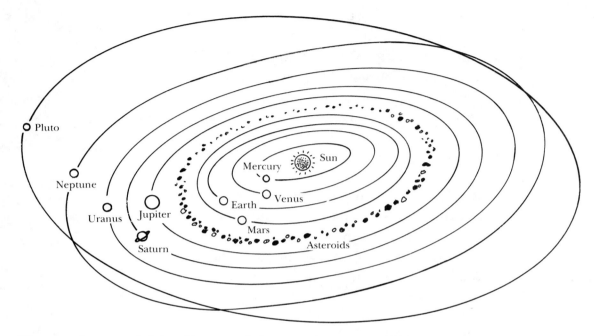

The nine planets and their orbits around the sun.

Sometimes you can see planets in the night sky. But to recognize them you have to know how to tell the difference between planets and stars. Stars produce their own light. Planets shine by reflecting light from the sun. Planets do not twinkle like stars. Instead, they shine with a steadier light. Try searching for a planet on a clear night.

Science Practice
1. What galaxy is the Earth part of?
2. What force holds the planets close to the sun?
3. How long does it take for the Earth to make one complete orbit of the sun?

The Inner Planets

Mercury is the planet closest to the sun. The daytime temperatures on Mercury are hot enough to melt lead. But at night it gets very cold there. Life as we know it could not exist on Mercury. Living things could not stand those temperature differences.

Venus is the second planet from the sun. The planet's surface is hidden from view by thick clouds of gases. These gases trap the sun's heat around the planet. This makes Venus a very, very hot place. Venus is even hotter than Mercury, though Venus is farther away from the sun.

The Earth is the third planet from the sun. It is roughly the same size as Venus. The water on Earth makes it a good place for living things. The oxygen, nitrogen, and carbon dioxide in our atmosphere are also necessary to life.

A **satellite** is a body that orbits a planet. The Earth's moon is an example of a natural satellite. The Earth's moon makes one complete orbit around the Earth every 28 days.

Mars is the fourth planet from the sun. It is known as the "red planet" because of the color of its surface. Mars has two tiny moons that revolve around it. Conditions on the surface of Mars are more like those on Earth than on any other planet. But so far no life

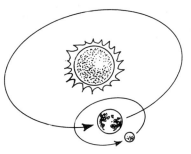

As the Earth orbits the sun, the moon orbits the Earth.

Chapter Twenty-Eight 325

has been discovered on Mars. The plants and animals of Earth could not live on Mars. There is only a trace of oxygen and water vapor in the Martian atmosphere. And the average temperatures on Mars are colder than the coldest places on Earth.

People in Science: Mae Jemison

On September 12, 1992, Dr. Mae Jemison became the first African-American woman to travel in space. She had dreamed of being an astronaut since she was a little girl growing up on Chicago's South Side. Jemison became a doctor first, however, before changing careers to become an astronaut.

Jemison's 8-day flight aboard the Space Shuttle *Endeavor* was very successful. She conducted experiments to find out more about motion sickness in space. The other astronauts on her flight took medicine to avoid motion sickness. But Jemison did not. Instead, she used the power of her mind and positive thinking to keep from getting motion sickness. She had practiced these techniques for months before the flight. Jemison also designed an experiment that looked at the effects of the space environment on bone cells.

Jemison enjoyed the whole flight. She says that when the *Endeavor* launched, she had a big smile on her face. She'd wanted to get closer to the stars for a long, long time. In March of 1993, Jemison resigned from NASA to pursue other interests in the fields of health care and education.

Mae Jemison

The Outer Planets

The outer planets are Jupiter, Saturn, Uranus, Neptune, and Pluto. These planets, except for Pluto, are much larger than the inner planets. Each outer planet has at least one moon. And each of these planets, except Pluto, has rings.

Jupiter, the fifth planet from the sun, is the largest planet in the solar system. It is about 11 times the size of the Earth. Jupiter has 16 moons. One of the moons, Io, has active volcanoes. Because of uneven heating in Jupiter's atmosphere, the planet has violent electrical storms. There is also a permanent hurricane on Jupiter called the "Great Red Spot." This area is about 18,000 miles long.

Saturn, sixth from the sun, is a very beautiful planet. It is surrounded by seven colorful rings. These rings are made up of tiny particles of ice and cosmic dust. The rings of Jupiter, Uranus, and Neptune are very faint and were only recently discovered. But Saturn's rings are easily seen from the Earth. They were discovered in the early 1600's by Galileo. In addition to the rings, Saturn is known to have at least 20 moons.

The last three planets in the solar system were unknown to early astronomers. *Uranus* was discovered in 1781. It is made up of ice and liquid hydrogen surrounding a solid core. Uranus has 15 known satellites. *Neptune*, the eighth planet from the sun, was discovered in 1846. It travels in a nearly circular orbit around the sun. Each orbit takes 165 years. Neptune has eight known satellites. A 7,900-mile-long blue spot, similar to Jupiter's Great Red Spot, has been discovered on Neptune.

Some scientists think other planets may yet be discovered some day.

Pluto is the farthest planet from the sun. It is so far from the sun that its surface is always in darkness. Some astronomers think Pluto may be an escaped moon of Neptune. Its orbit actually crosses inside of Neptune's on its path around the sun. Pluto has one satellite.

People Explore Space

In 1957 the first man-made satellite circled the Earth. This satellite, called Sputnik, was sent into space by the Soviet Union. Today many satellites circle the Earth. These satellites gather all kinds of information. They help with weather reports, with TV and telephone communication systems, and with navigation.

On July 20, 1969, a spacecraft called Apollo 11 landed on the moon. Two astronauts stepped out of the spacecraft. Their names were Neil Armstrong and Buzz Aldrin. They were the first people to walk on the moon.

In 1976, two spacecraft, Viking 1 and Viking 2, landed on different parts of Mars. Many photographs of the rocky deserts of Mars' surface were taken. Some pictures show what look like dry river beds. Scientists wonder if Mars used to have large amounts of surface water. If it did, it could have had life, too. In fact, scientists believe that water may be frozen in the polar caps of Mars or beneath the surface. There are even some scientists who think that some form of life may still exist on Mars.

Astronauts Buzz Aldrin and Neil Armstrong were the first people to walk on the moon.

Asteroids, Meteors, and Comets

Between the orbits of Mars and Jupiter is a belt of rocky objects. These small rocky objects are called **asteroids**. Like planets, asteroids orbit the sun. Asteroids are rich in minerals. One day people may mine for minerals in the asteroid belt.

Amazing Science Fact

Astronomers are not sure where asteroids are from or how they were formed. Asteroids may be the remains of a planet that broke up. Some people believe that asteroids are left-over materials from the formation of the solar system.

Fragments from space sometimes enter the Earth's atmosphere. Most of these fragments don't make it to the Earth. As they fall through the atmosphere they become very hot and burn up. The heat is created by the friction of the fragment against the air. As they burn, they make a bright streak of light. Such fragments are called **meteors**.

Sometimes rocky fragments from space make it all the way to the Earth's surface without burning up. These fragments are called **meteorites**. A few hundred meteorites are found on Earth each year. Many are no bigger than baseballs. They are made of many different stony materials mixed with iron particles.

Have you ever seen a shooting star? Shooting stars are meteors that burn as they enter the Earth's atmosphere.

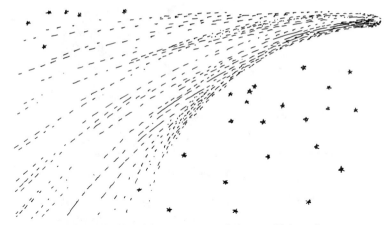
Comets are huge balls of ice, gases, and dust orbiting the sun.

One very famous comet is Halley's Comet. It passes close enough to the sun to glow about once every 76 years.

Comets are bodies of ice, gases, and bits of dust that orbit the sun. When comets pass close to the sun their gases glow, forming a visible tail. Most comets are only about a mile in diameter. But their tails can be 100,000,000 miles long.

On the Cutting Edge
On April 24, 1990, the most powerful telescope ever built, the Hubble Space Telescope, was launched into space. Forty years had been spent designing, raising money, and building the telescope. The telescope is about the size of a boxcar. It weighs 13 tons (11,600 kg) and cost 1.55 billion dollars. It is the heaviest and most expensive scientific spacecraft yet lifted into orbit.

The Earth's atmosphere bends light from the stars and galaxies. So the pictures of space that we get from Earth are blurred. The Hubble was built to orbit Earth about 360 miles (580 km) above the surface. It would take pictures *above* the atmosphere. Scientists expected clear pictures. They hoped the Hubble Space Telescope would help them learn the size and age of the universe. And it might. However, there have been some serious problems.

First, the mirror was made the wrong shape by mistake. It does not form images as sharply as expected. Also, the craft's wings cause it to wiggle. This wiggling also blurs the pictures. Sometimes the telescope misses its target altogether.

Even so, the powerful telescope has taken pictures of space that are far better than any ever taken before. It has observed more than 300 galaxies four billion light-years away. Astronomers think they are seeing the solar system as it looked when it first began.

A light-year is the distance light travels in a year, or 5.88 trillion miles.

Chapter Review

Chapter Summary

- The universe is all the matter, energy, and space in existence. A galaxy is a group of stars, dust, and gases held together by gravity. The sun is a star in the galaxy called the Milky Way.

- The solar system is made up of the sun and the nine planets that circle the sun. The huge mass of the sun has a strong gravitational pull on the planets. The inner planets are Mercury, Venus, Earth, and Mars. The outer planets are Jupiter, Uranus, Saturn, Neptune, and Pluto.

- Asteroids are rocky fragments that circle the sun between the orbits of Jupiter and Mars. When a rocky fragment falls to Earth before burning up, it is called a meteorite. A rocky fragment that burns up while passing through the atmosphere is called a meteor.

- A comet is a ball of ice, frozen gases, and dust. Comets have elliptical orbits around the sun. When they pass close to the sun their gases glow. They then make a trail across the sky.

Chapter Quiz

Answer the following questions on a separate sheet of paper.

1. What is the sun?
2. What galaxy is the sun in?
3. Name at least three kinds of objects found in the solar system.
4. Which planets in the solar system have rings?
5. What determines the length of a "day" on a planet? How long is a day on Earth?
6. What is the natural satellite that orbits Earth called?
7. Name two uses for artificial satellites.
8. Why do astronomers think there may have been water on Mars?
9. What is another name for a "shooting star"?
10. What is a meteorite?

Mad Scientist Challenge: Find Yourself in the Universe

Look at the following list of words. On a separate sheet of paper write the words in order of biggest place to smallest place.

American continent

solar system

galaxy

your street (write its name)

your state (write its name)

the Earth

universe

your town or city (write its name)

United States

Chapter Twenty-Eight

Unit 4 Review

Answer the following questions on a separate sheet of paper.

1. How many time zones are there in the continental United States? If it is two o'clock in the afternoon in New York, what time is it in San Francisco?
2. What substance covers most of the Earth's surface?
3. Explain the theory of plate tectonics.
4. What can happen when two of the Earth's plates collide?
5. List the five layers of the Earth's atmosphere. Start with the layer closest to the Earth.
6. What causes most winds?
7. Cumulonimbus clouds are often accompanied by what kind of storm?
8. What are the names of the four geological eras? Name the main characteristics of each era.
9. What causes tides? What conditions create neap tides? What conditions create spring tides?
10. Name the nine planets of the solar system in the order of their distance from the sun. Begin with the planet closest to the sun.

Appendix

Glossary

Appendix A: Careers in Science

Appendix B: Hiring Institutions

Appendix C: Metric Conversion Chart

Appendix D: Table of Chemical Elements

Appendix E: The Five Food Groups

Appendix F: Minerals

Appendix G: Vitamins

Index

Glossary

air pressure   the weight of the gases pressing down on the Earth

algae   (singular, alga) plant-like protists.

amphibian   an organism that lives part of its life in water and part on land, such as a frog

amplitude   the height of a wave

appendage   a body part that sticks out, such as a wing, feeler, arm, or leg

arteries   blood vessels that carry blood away from the heart

arthropod   an animal with a shell, jointed appendages, and body divided into segments

asteroids   rocky fragments that circle the sun between the orbits of Jupiter and Mars

astronomers   scientists who study the universe beyond the Earth's atmosphere

atmosphere   the gas that surrounds the Earth

atoms   tiny parts into which all things on Earth can be broken down

axis   an imaginary line running from one pole, through the center of a planet, to the other pole

bacteria   (singular, bacterium) simple, one-celled organisms that are visible only through a microscope; a kind of moneran

barometer   an instrument that measures air pressure

battery   a device that changes chemical energy into electrical energy

biology   the science that studies life

biologist   a scientist who studies the habits and growth of living organisms

bit   the smallest unit of information used by a computer; represented by a 0 or a 1

botany   the scientific study of plants

breeding   producing offspring; raising plants or animals, especially to get new or better kinds

byte   a string of eight bits standing for a single character

capillaries   tiny blood vessels that connect arteries and veins

carbohydrates   sugars in food that give people energy

carbon dioxide   a gas made of carbon and oxygen molecules

cell   the tiny basic unit of which all living things are made

Cenozoic era   the current geological era, which began 65,000,000 years ago

centripetal force   a force that causes objects to move in a curved path

characteristics the qualities or features of a person or thing

chemical energy energy stored in molecules

chemistry the scientific study of what substances are made of and how they can change when they combine with other substances

chlorophyll the green coloring in chloroplasts that traps sunlight

chloroplasts the special parts of a plant cell that help make food from sunlight

chromosomes thread-like parts of a cell nucleus made up of DNA and genes

circuit an unbroken path along which an electrical current flows

circulatory system the system that carries blood through the body, delivering food and oxygen to all parts of the body

cirrus clouds high altitude clouds made of ice crystals

classification the way in which biologists group organisms by type

climate the average weather in a region over many years

cold-blooded having a body temperature that changes with the temperature of the environment

comet a ball of ice, dust and gases that orbits the sun

community all the organisms that live in one habitat

compound a substance that is made when two or more elements join chemically

conduction the bumping of molecules that moves heat through matter

conductor a material through which electricity travels well

condensation the process by which gas turns into a liquid

conservation the wise and careful use of our natural resources

consumers organisms that eat other organisms

continents the seven major land masses on Earth: Africa, North America, South America, Asia, Europe, Antarctica, and Australia

continental shelf the gently sloping part of the ocean floor that begins at the edge of continents

continental slope the steep slope of the ocean floor between the continental shelf and the ocean basin

convection the transfer of heat within a gas or liquid by the movement of warmer particles

core the center of the Earth

crossbreeding combining the sex cells of organisms with different traits to create new traits

crust the thin, outer layer of the Earth

cumulonimbus a tall, thick, white cloud with a dark base, often known as a thunderhead

cumulus clouds large clouds that have flat bases and rounded masses piled on top

cyclone an area of low air pressure with circular wind motion

cytoplasm the watery substance in a cell

data information, facts, numbers, or letters processed by a computer

decomposers organisms that eat dead matter

density the amount of matter per unit volume

dew point the temperature at which water vapor turns to liquid water

digest to break down food inside the body or cell into a form that can be used

digestive system the system that breaks down food so that body cells can use it

discharge the letting go of extra electrons

disease sickness; a condition in which the body is not functioning properly

disk drive the machine that reads data on diskettes and records new data onto diskettes

diskettes disks made of magnetic tape and used to store data; they can be removed from the computer

DNA molecules in the nuclei of cells that control many of the characteristics of living things

ecology the study of how all living things on Earth depend on one another

ecosystem a system that is formed by the interaction of a community of organisms and its environment

effort force the force applied when doing work

electrical energy energy produced and carried by the electrons in a substance

electrical insulator a material through which electricity does not travel

electrons particles with negative electrical charges, surrounding the nucleus of an atom

elements the basic substances of which all matter is made

energy the ability to do work

environment all the things around you

enzymes substances that cause chemicals to change form in the body

erosion the wearing away of soil by wind and water

equator an imaginary line that circles the Earth halfway between the two poles; the 0 degree line of latitude

esophagus the tube that carries food from the mouth to the stomach

evaporation the process by which heat changes water to water vapor

evolution changes in a species over time

evolve to change over time

experiments tests that are used to discover or prove something

extinct no longer living on the Earth; referring to a species that has died out

fats nutrients in foods that supply the body with fuel or energy

feces solid wastes expelled from the body

fertilization the joining of a sperm cell with an egg cell

fetus a young mammal after fertilization of the egg but before birth

field a profession or special area of interest

food chain a group of organisms, each of which is dependent on another for food

force any push or pull on an object

fossil fuels fuel products made from plant and animal remains, such as coal, petroleum, and natural gas

fossils the remains of organisms that lived long ago

frequency the number of wave cycles that pass through a point in one second

friction a force that resists motion

front the place where two air masses of different temperatures meet

fulcrum the support on which a lever turns

fungi (singular, fungus) organisms that cannot move around like animals and do not have chloroplasts like plants. Mushrooms, molds, and yeasts are fungi.

fuse a weak link in an electrical circuit that is designed to break the circuit if it gets too hot

gas matter without a definite shape or a definite volume

gene a part of a chromosome that controls the development of individual traits

genetics the study of how life characteristics are passed along to offspring

generator a machine that turns mechanical or heat energy into electrical energy

geological eras periods of time in the Earth's history

geologist a scientist who studies the history and structure of the Earth, especially as recorded in rocks

geothermal energy energy from hot water trapped under layers of rock deep in the Earth

germination the process by which a young plant breaks out of its seed

geyser a hot spring from which steam and hot water shoot into the air

gills organs that allow fish to get oxygen from water

glacier a large slow-moving field of ice

globe a map of the world that is round like a ball

gram a basic unit of weight in the metric system, equal to 0.035 ounces

gravity the force of attraction between any two objects that have mass

habitat the place where an organism lives

heat energy energy produced by the motion of molecules

heredity the passing of traits from parents to offspring

hormones substances produced by certain glands in the body. These substances are circulated in the blood

host an animal on or in which parasites live

humidity the amount of moisture in the air

hurricane a stormy tropical cyclone over the North Atlantic Ocean

hybrids the offspring of crossbreeding

hydroelectric energy electrical energy produced by the movement of water

igneous rocks rocks formed from magma

inclined plane a slanted surface used for raising objects from one level to another

inertia the tendency to stay at rest or in motion unless acted on by a force

infinity space and time without end

input data entered into the computer

insulator a material that does not conduct heat well

invertebrates animals without backbones

ionosphere a layer in the upper atmosphere that begins in the mesosphere and extends upward through the thermosphere

keyboard the input device used to type information into the computer, much like a typewriter keyboard

kinetic energy energy in movement

kingdoms the five main groups in biological classification

lava what magma is called once it reaches the Earth's surface

lever a simple machine made of a bar that turns on a support

life span the amount of time an organism is likely to live

light energy energy produced by the motion of waves of light

lines of latitude lines drawn east and west on a map to help locate places

lines of longitude lines drawn north and south on a map to help locate places

liquid matter with a definite volume but no definite shape

liter a basic unit of volume in the metric system, equal to about 1.05 quarts of liquid

load the object moved through a distance in work

lubricants substances that reduce friction between moving parts of machines

machine any device that can change the speed, direction, or amount of a force

magma melted rock squeezed up from the Earth's mantle

magnet a stone, piece of metal, or any solid substance that attracts iron or steel

magnetic field the area around a magnet that exerts a magnetic force

magnetic tape a plastic tape covered with magnetic material; used for recording data

mammals animals that have hair on their bodies and give birth to live young. Female mammals provide milk for their young.

mantle the layer of the Earth between the crust and the core.

mass the amount of matter in something

measurement the size, quantity, or amount of something

mechanical advantage the number of times a machine multiplies an effort force

mechanical energy energy produced by the moving parts of a machine

membrane the protective covering that holds a cell together

menopause the time in a woman's life when menstruation stops

menstruation the monthly shedding of blood from the uterus when an egg has not been fertilized

mesosphere the third layer in the atmosphere

Mesozoic era the geological era beginning 225,000,000 years ago and lasting 160,000,000 years

metamorphic rocks rocks formed when igneous or sedimentary rocks change under very high temperatures or pressure

meteor a rocky fragment from space that burns up when it enters the Earth's atmosphere

meteorology the scientific study of the Earth's atmosphere

meter the standard unit of length in the metric system, equal to 39.4 inches, or slightly more than 3 1/4 feet

metric units the standard units of measurement in the metric system, based on the number ten and multiples of ten

microbiology the study of organisms too small to be seen with the unaided eye

microscope a machine for viewing objects that are too small to be seen by the naked eye

mid-ocean ridge an underwater mountain chain that runs down the middle of the ocean floor

migrate to move at regular times of year from one region to another for food, mating, or warmer temperatures

minerals substances found in non-living things; people need minerals in their diets in order to stay healthy

mixture a substance that is made when two or more elements mix but do not join chemically

molecule two or more atoms joined together

monera (singular, moneran) tiny organisms that have some nucleic materials, but no true nuclei, in their cells, such as bacteria

monitor the part of the computer that has a screen showing what is going on in the computer

motion a change in the position or place of an object

mutation a change in the genetic code of an organism

natural resources substances found in nature that are useful to humans

natural selection the way that those organisms best suited to their environments survive and pass their helpful traits along to offspring

neurons cells throughout the body that carry signals to and from the brain (also called nerve cells)

neutrons particles with no electrical charge, found within the nucleus of an atom

nicotine a substance in cigarettes that can cause heart disease and lung cancer

nuclear energy energy stored in the nucleus of an atom

nuclear fission the splitting of atomic nuclei, resulting in great energy release

nuclear fusion the joining of atomic nuclei, resulting in great energy release

nucleus (plural, nuclei) the part of the cell that controls all the other parts

nutrients substances in foods that body cells need

observations careful studies of something, especially for scientific purposes

occluded front the front that forms when a cold air mass overtakes a warm air mass

ocean basin the bottom of the ocean floor

oceanographer a scientist who studies the oceans

organ body tissues that form a working unit, such as a heart or kidney

organism a living thing

output processed data that comes out of the computer

ovaries the female organs that make egg cells and hormones

ozone a form of oxygen in a thin layer within the stratosphere

Paleozoic era the geological era beginning 570,000,000 years ago and lasting 346,000,000 years

parasite an organism that lives on or in another organism

petals the colored outer parts of a flower that help protect its inner parts and that attract insects

photosynthesis the process by which plants make sugar using sunlight, water, chlorophyll, and carbon dioxide

physics the scientific study of energy and how it interacts with matter

pistil the female part of a flower

plant ovary the fruit of a plant that grows around the seed

plasma the liquid part of blood

plate a large piece of the Earth's crust

platelets solids in the blood that help stop bleeding at an injury

plate tectonics the theory that the Earth's crust is made of plates that slowly shift position

pollen the powdery dust on the ends of stamens that holds plant sperm cells

pollination the transfer of pollen from the stamen to the pistil of a plant

population the group of one species living in a certain place

potential energy stored energy

Precambrian era the geological era beginning 4,500,000,000 years ago and lasting nearly four billion years

precipitaton any moistutre that falls from the atmosphere

prime meridian the 0 degree of longitude

prism a triangular-shaped object made of clear glass. It can break up a ray of white light into the colors of the rainbow.

process a series of steps for making or doing something

processing working with data on a computer

producers organisms that make their own food

program a set of coded instructions that leads a computer through certain tasks

properties qualities of matter such as color, shape, odor, and hardness

proteins nutrients in foods that build body tissues

protists tiny one-celled organisms that are neither plants nor animals but that often have characteristics of both

protozoans (singular, protozoan) animal-like protists

puberty the time in life when the reproductive organs develop

pulley a simple machine made of a wheel that turns on an axle. A rope or chain can be pulled around a groove in the wheel.

radiation energy that can move through a vacuum

radioactive dating finding the age of rocks by measuring the decay of radioactive elements

recycle to use something over and over again

red blood cells blood cells that carry oxygen and carbon dioxide

reflection light bouncing off an object

refraction the bending of light rays when they pass from one material to another

reproduce to make more of one's kind

resistance force the force that must be overcome in work

respiration the way a cell gets energy by mixing food and oxygen

respiratory system the system that takes in oxygen and combines it with food to produce energy

respond to act in return or in answer to something

salinity a measure of saltiness

satellite an object that revolves around a plant. Both moons and man-made objects put into orbit are satellites.

science the study of nature and the universe, based on facts that are learned from observation and experiment

screw a simple machine made of an inclined plane wrapped around the length of a nail

sedimentary rocks rocks formed by many different rock particles "cementing" together

seismic sea waves giant waves caused by earthquakes on the ocean floor

solar collector a tool used for collecting sunlight and transforming it into heat energy

solar energy energy produced by the sun

solar system the sun and all the planets that revolve around it

solid matter with a definite shape and volume

solution a kind of mixture in which one substance is dissolved into another

species the smallest groups in biological classification

spectrum the rainbow-like band of color that is seen when white light is refracted

sperm male sex cells

stamens the male parts of a flower

static electricity the electricity created when objects with opposite charges are attracted to each other

stratosphere a layer in the atmosphere that begins about seven miles up; the second layer in the atmosphere

stratus clouds low foglike clouds that lie in a flat layer over a wide area

structure the way an organism is put together

system a group of organs working together

technology the application of scientific and industrial skills to practical use

tendons tough bands of tissue that attach muscles to bones

testes the male organs that make sperm cells and hormones

theory a group of ideas or principals that explain why something happens

thermosphere the fifth layer in the atmosphere

tides the rising and falling of sea level, caused by the gravitational pull of the moon and the sun

tissue a group of cells that all do the same job

tornado a cyclone with violent whirling winds forming a funnel-shaped cloud that extends downward from a cumulonimbus cloud

traits characteristics, which may be inherited, that identify organisms as individuals

trench a deep, long valley in the ocean floor

troposphere the first layer in the atmosphere; the lower atmosphere in which most weather takes place

turbine a machine driven by the force of a moving fluid

typhoon a stormy tropical cyclone over the Pacific Ocean

undertow a current of water that washes back out to sea under incoming waves

unit a fixed amount or quantity that is used as a standard of measurement

universe everything that exists, including the Earth, sun, planets, stars and outer space

uterus the female organ in which a fertilized egg develops into a baby

vaccine a substance injected into a person to keep him or her from getting a certain disease

vacuoles openings in a cell that store food, water, or wastes

vacuum the absence of matter

veins blood vessels that return blood to the heart

vertebrates animals with backbones

virus a microscopic "organism" that causes diseases and is missing some cell parts; viruses can grow and reproduce only in certain living cells

vitamins substances found in many foods; people need vitamins in their diets in order to stay healthy

volcano a hole in the Earth's surface through which magma pours from the mantle

wastes the leftover matter a cell or body does not need after it uses food for energy

wavelength the distance from the crest of one wave to the crest of the next

weathering the process that breaks down rocks and minerals

wedge a simple machine made of a tapering piece of wood, metal, or other material

weight the measure of the force of gravity

wheel and axle a simple machine made of a wheel attached to a shaft

white blood cells blood cells that fight bacteria

work the force moving something through a distance

zoology the scientific study of animals

Appendix A

Careers in Science

This is a partial list of careers in science-related fields.

Air pollution analysis
Anatomy
Animal breeding
Animal ecology
Anthropology
Aquarium work
Astronomy
Biochemistry
Biological photography
Biomedical engineering
Biophysics
Botany
Chemistry
Chiropractic
Computer programming
Criminology
Dentistry
Engineering
Entomology
Environmental engineering
Environmental law
Environmental science
Farming
Food chemistry
Food-and-drug inspection
Forestry
Gardening
Genetics
Geology
Hospital work
Hydrology
Laboratory testing
Landscape contracting
Marine biology
Medical illustration
Medicine
Meteorology
Microbiology
Mineralogy
Nursing
Nursery management
Nutrition
Occupation-safety-and-health inspection
Oceanography
Optometry
Paleontology
Parasitology
Park maintenance
Pathology
Pharmaceutical science
Physical therapy
Plant pathology
Podiatry
Psychology
Public-health education
Ranching
Range management
Safety inspection
Science editing
Science writing
Scientific illustration
Seismology
Soil conservation
Taxidermy
Teaching
Technical writing
Veterinary science
Water-quality analysis
Weather observation
Wildlife management
Zoology

Appendix B

Hiring Institutions

This is a partial list of the sorts of institutions that hire people for science-related jobs.

- Advertising agencies
- Aquariums
- Arboretums
- Beverage companies
- Botanical gardens
- Business corporations
- Chemical industries
- Colleges and schools
- Consulting firms
- Cosmetic companies
- Doctors' offices
- Educational institutions
- Food processors
- Government agencies
    - Agricultural Department
    - Energy Department
    - Environmental Protection Agency
    - Fish and Wildlife Service
    - Health and Human Services Department
    - National Institute of Health
    - National Science Foundation
    - Patent Office
    - Peace Corps
    - Vista
- Hatcheries
- Hospitals
- Import/export companies
- Libraries, medical and technical
- Lumber companies
- Manufacturing industries
- Medical clinics
- Medical laboratories
- Medical-supply companies
- Museums
- National and state parks
- Nurseries
- Petroleum companies
- Pharmaceutical companies
- Professional and technical journals
- Publishers
- Research and development firms
- Textile manufacturers
- Utility companies
- Zoological parks

Appendix C

Metric Conversion Chart

Length
1 millimeter = .03937 inch
1 centimeter = .3937 inch
1 decimeter = 3.937 inches
1 meter = 39.37 inches = 3.281 feet = 1.0936 yards
1 kilometer = 1,093.6 yards = .6214 mile
1 inch = 2.54 centimeters
1 foot = 30.48 centimeters = .3048 meter
1 yard = 91.44 centimeters = .9144 meter
1 mile = 160,933 centimeters = 1,609.33 meters = 1.60933 kilometers

Area
1 square millimeter = .00155 square inch
1 square centimeter = .155 square inch
1 square decimeter = 1.55 square inches
1 square meter = 1,550 square inches = 10.764 square feet = 1.196 square yards
1 square inch = 6.4516 square centimeters
1 square foot = 929.03 square centimeters
1 square yard = .8361 square meters

Volume
1 cubic inch = 16.387 cubic centimeters
1 cubic foot = 28,317 cubic centimeters = .0283 cubic meter
1 cubic yard = 764,553 cubic centimeters = .7646 cubic meter
1 fluid ounce = 29.573 milliliters
1 quart = .9463 liter
1 gallon = 3.7853 liters
1 liter = 1.0567 quarts

Weight
1 gram = .035 ounce
1 kilogram = 2.2 pounds
1 metric ton = 2,200 pounds

Appendix D

Table of Chemical Elements

Element	Symbol	Atomic Number (number of protons)
actinium	Ac	89
aluminum	Al	13
americium	Am	95
antimony	Sb	51
argon	Ar	18
arsenic	As	33
astatine	At	85
barium	Ba	56
berkelium	Bk	97
beryllium	Be	4
bismuth	Bi	83
boron	B	5
bromine	Br	35
cadmium	Cd	48
calcium	Ca	20
californium	Cf	98
carbon	C	6
cerium	Ce	58
cesium	Cs	55
chlorine	Cl	17
chromium	Cr	24
cobalt	Co	27
copper	Cu	29
curium	Cm	96
dysprosium	Dy	66
einsteinium	Es	99
erbium	Er	68
europium	Eu	63
fermium	Fm	100
fluorine	F	9
francium	Fr	87

Table of Chemical Elements (continued)

Element	Symbol	Atomic Number (number of protons)
gadolinium	Gd	64
gallium	Ga	31
germanium	Ge	32
gold	Au	79
hafnium	Hf	72
helium	He	2
holmium	Ho	67
hydrogen	H	1
indium	In	49
iodine	I	53
iridium	Ir	77
iron	Fe	26
krypton	Kr	36
lanthanum	La	57
lawrencium	Lr	103
lead	Pb	82
lithium	Li	3
lutetium	Lu	71
magnesium	Mg	12
manganese	Mn	25
mendelevium	Md	101
mercury	Hg	80
molybdenum	Mo	42
neodymium	Nd	60
neon	Ne	10
neptunium	Np	93
nickel	Ni	28
niobium	Nb	41
nitrogen	N	7
nobelium	No	102
osmium	Os	76

Table of Chemical Elements (continued)

Element	Symbol	Atomic Number (number of protons)
oxygen	O	8
palladium	Pd	46
phosphorus	P	15
platinum	Pt	78
plutonium	Pu	94
polonium	Po	84
potassium	K	19
praseodymium	Pr	59
promethium	Pm	61
protactinium	Pa	91
radium	Ra	88
radon	Rn	86
rhenium	Re	75
rhodium	Rh	45
rubidium	Rb	37
ruthenium	Ru	44
samarium	Sm	62
scandium	Sc	21
selenium	Se	34
silicon	Si	14
silver	Ag	47
sodium	Na	11
strontium	Sr	38
sulfur	S	16
tantalum	Ta	73
technetium	Tc	43
tellurium	Te	52
terbium	Tb	65
thallium	Tl	81
thorium	Th	90

Table of Chemical Elements (continued)

Element	Symbol	Atomic Number (number of protons)
thulium	Tm	69
tin	Sn	50
titanium	Ti	22
tungsten	W	74
uranium	U	92
vanadium	V	23
xenon	Xe	54
ytterbium	Yb	70
yttrium	Y	39
zinc	Zn	30
zirconium	Zr	40

Note: Three new elements have been discovered in the last 25 years. However, these elements have not yet been officially approved or named.

Appendix E

The Five Food Groups

Food Group	Sample Foods	Main Nutrient Contributions
Meat and meat substitutes	Beef, pork, lamb, fish, poultry, eggs, nuts, legumes	Protein, iron, riboflavin, niacin, thiamine
Dairy products	Milk, buttermilk, yogurt, cheese, cottage cheese, soy milk, ice cream	Calcium, protein, riboflavin, thiamine
Fruits	apples, bananas, grapes, kiwis, melons, oranges pears, pineapples	Vitamin A, Vitamin C, thiamine, iron, riboflavin
Vegetables	broccoli, celery carrots, cabbage lettuce, peas potatoes, radish	Vitamin A, Vitamin C, thiamine iron, riboflavin
Grains (bread and cereal products)	All whole-grain and enriched flours and products	Riboflavin, iron, thiamine, niacin

Appendix F

Minerals

Mineral	Source	Body Function
Calcium	Milk, vegetables, meats, dried fruits, whole-grain cereals	Healthy bones and teeth; blood clotting; muscle spasms
Iodine	Saltwater fish, shellfish, iodized salt	Functioning of thyroid gland; regulation of use of energy in cells
Iron	Liver, meats, eggs, nuts, dried fruits, green leafy vegetables	Formation of red blood cells
Magnesium	Milk, meats, whole-grain cereals, peas, beans, nuts, vegetables	Normal muscle and nerve action; regulation of body temperature; building strong bones
Phosphorus	Milk, meat, fish, poultry, nuts, vegetables, whole-grain cereals	Bone and teeth formation; nerve and muscle function; energy production
Potassium and Sodium	Most foods, table salt (sodium)	Blood and cell functions; balance of fluids in tissue

Appendix G

Vitamins

Vitamin	Source	Body Function
A	Milk, butter, margarine, liver, leafy green and yellow vegetables	Growth; health of eyes and skin
B_1 (thiamine)	Cereals, bread, lean meat, liver, milk, green vegetables, fish	Growth; working of the heart, nerves, and muscles
B_2 (riboflavin)	Meat, soybeans, milk, green vegetables, eggs, poultry	Healthy skin; prevents sensitivity of eyes to light, building and maintaining tissue
B_{12}	Green vegetables, liver	Prevents anemia
Niacin	Meat, poultry, fish, peanut butter, potatoes, whole grain	Prevents pellagra; healthy nervous system
C	Citrus fruits, leafy vegetables, fruits	Prevents scurvy; healthy blood cells, strong body cells
D	Fish-liver oil, liver, fortified milk, eggs	Prevents rickets; aids metabolism of calcium and phosphorus

Vitamins (continued)	Vitamin	Source	Body Function
	E	Vegetable oils, wheat germ, whole grains, lettuce	Prevents infertility, muscular dystrophy
	K	Leafy vegetables	Aids blood clotting

Index

A
Acid test, 157
AIDS, 132
Air masses, 284
Air pollution, 273
Air pressure, 274–275
Alchemists, 169
Aldrin, Buzz, 328
Algae, 54, 67, 316
Almanacs, 292–293
Amphibians, 69
Amplitude, 202
Animal kingdom, 60–73
Animals
    breeding of, 93
    cells of, 43
    cold-blooded, 68
    definition of, 62
    food and, 30
    group names, 73
    movement and, 30
    reproduction and, 31
    warm-blooded, 72
Antennae, 66
Anus, 65, 123
Apollo 11, 328
Appendages, jointed, 66
Arctic ice sheet, 303
Area, metric units of, 18
Aristotle, 4–5, 53
Armstrong, Neil, 328
Arteries, 121
Arthropods, 66
Asteroids, 329
Astronomers, 321
Athlete's foot, 56
Atmosphere
    definition of, 272
    heating of, 275
    layers of, 272–274
    water in, 278–279
Atoms, 4, 38
    electricity and, 212
    magnetism and, 219
    structure of, 155–156
Attraction
    electrical, 213
    magnetic, 219
Axis
    of Earth, 251–252
    of planets, 323

B
Bacteria, 55
    disease and, 130
    insulin and, 92
    penicillin and, 21
Banneker, Benjamin, 292–293
Barometers, 275
Batteries, 216
    solar, 227
Beams, light, 202
Bees, 66, 82
Berriozabal, Manuel, 241
Biology. *See* Life science
Birds, 71–72
Bits, 236
Black Death. *See* Plague
Black widow spider, 67
Blind, systems of reading for, 6
Blood, 120–122
    respiration and, 125
Blood cells
    red, 121
    white, 121–122, 130
Blood pressure, high, 122
Blue-green algae, 55
Body image, 135
Body segments, 66
Bones, 113
Botany, 28
Botulism, 56
Braille, 6
Braille, Louis, 6
Brain, human, 108, 112
Bread, 57
Breakers, 312
Breathing. *See* Respiration
Breeding, 93
    genetics and, 88–90
Bronchi, 124
Bruises, 121
Bytes, 236

Index   361

C

Calcium, 113
Cancer, 135
Capillaries, 121
Carbohydrates, 133
Carbon, 155
Carbon dioxide
    humans and, 125
    plants and, 79–80
Cassava, 7
Cells, 38
    animals and, 62
    blood, 121–122
    definition of, 39
    energy and, 42
    main parts of, 40
    plant vs. animal, 43
    specialized, 62–63
Cell walls, 43
Cenozoic era, 302
Centimeter, 18
Centripetal force, 177
Characteristics, definition of, 29
Charge, electric, 156, 212
    lightning and, 207, 214
Chemical bonds, 158–159
Chemical energy, 165
Chemistry, 154
Chlorophyll, 79
Chloroplasts, 43, 56
Cholesterol, 134
Chromosomes, 90
    genes jumping around on, 103
Circuits, electrical, 217–218
Circulatory system, 109, 120–122
Cirrus clouds, 279
Classifications systems, 51–53
Climate
    definition of, 290–291
    types of, 291–292
Closed circuit, 218
Clouds, 279
    seeding of, 288
Coal, 301
Cold-blooded animals, 68
Cold fronts, 285
Color, 205–206
Comets, 330
Communities, 140
Compact disk players, 6
Competition, between organisms, 101
Compound machines, 191
Compounds, 158–159
Computers
    machines and, 193
    maps and, 254
    parts of, 234–236
    uses of, 239–241
Condensation, 167
Conduction, 199–200
Conductors
    of electricity, 215
    of heat, 199
Conservation, 146–147
Consumers, 142–143
Continental shelves, 315
Continental slopes, 315
Continents, 248
Convection, 200
Convection currents
    ocean and, 310–311
    winds and, 275–276
Core, Earth, 250
Crest, wave, 202
Crick, Francis, 45
Crocodiles, 71
Crop, 65
Crossbreeding, 89
Crust, Earth, 250
    movement of, 260–261
Crustaceans, 67
Cumulonimbus clouds, 285
Cumulus clouds, 279
Currents
    electrical, 215–216
ocean, 309–311
Cyberspace, 238
Cyclones, 286–287
Cytoplasm, 40

D

Dams, 228–229
Darwin, Charles, 100
Data, 234
    management of, 240
Dating, radioactive, 298
Dead Sea, 249, 309
Decomposers, 143
Density, 157

Dew point, 167, 278
Diabetes, 92
Diamonds, 155
Dieting, 135
Digestion, cells and, 42
Digestive system, 109, 122–123
Dinosaurs, 98, 301
   extinction of, 99
Discharge, electrical, 213
Disease
   definition of, 130
   insects and, 67
Disk drives, 236
Diskettes, 236
DNA
   evolution and, 100
   importance of, 44
   model of structure, 45
DuPont Company, 168

E

Earle, Sylvia, 317
Ears, 109–110
Earth
   age of, 249
   distance from sun, 27
   features of, 248–249
   gravity and, 174
   layers of, 350
   movement of, 251–252
   size of, 250
Earthquakes, 259–260, 263
Earth science, 8
Earthworms, 65
Ebbesmeyer, Curtis C., 16
Ecology, 29, 140
Ecosystems, 140
   change in, 141–142
Edison, Thomas, 192
Effort force, 184–185
Eggs
   genes and, 91
   human, 114
   plant, 81
Electrical energy, 165
Electricity
   charge, 156, 212
   circuits, 217–218
   currents, 215–216
   discharge, 213
   thunderheads and, 286
Electrons, 155–156
   electricity and, 212
   magnetism and, 219
Elements, chemical, 38, 155
   table of, 350–353
Energy
   cells and, 42
   chemical, 165
   definition of, 164
   electrical, 165
   forms of, 165
   geothermal, 229
   heat, 165, 167, 198–199
   hydroelectric, 228–229
   kinetic, 164, 198
   light, 165
   machines and, 186
   mechanical, 165
   nuclear, 166, 224–226
   from oceans, 316
   physical science and, 154
   potential, 164
   solar, 143, 227–228
   sources of, 143, 223–229
   tides and, 229
   volcanoes and, 229
   water and, 228–229
   waves, 201–202
   wind and, 229
Environment, 32–33
   traits and, 93
Enzymes, 122
Equator, 249
   currents and, 311
Erosion, 267
Esophagus, 122
Evaporation, 144, 167
Evolution
   definition of, 98
   DNA and, 100
   mutation and, 102–103
Exercise, 135
Experiments, 5, 14–15
Extinction, 98
Eyes, 109

F

Fats, 133
Feces, 123
Fertilization, 91
Fetus, 114
Fields
    career, 8–9, 347
    magnetic, 219
Filaments, 217
Fishes, 68–69
Fleming, Alexander, 21
Flowers, 81–82
Fluid friction, 176
Fog, 288
Food
    animals and, 62
    cells and, 42
    getting and using, 30
    natural selection and, 101
    plants and, 78–80
Food chains, 143
Food cycle, 142–143
Food groups, 354
Food webs, 143
Fool's Gold, 157
Force, definition of, 173–174
Forests, 80
Fossil fuels, 143, 223–224
Fossils, 98, 298
Franklin, Benjamin, 215
Fraud, scientific, 169, 186
Freezing, 167
Frequency of waves, 202
Friction, 176
    earthquakes and, 263
Frogs, medicine and, 69–70
Fronts, weather, 284
Fruit flies, 91
Fruits, 83
Fulcrums, 186–187
Fungus kingdom, 56–57
Fuses, 218

G

Galaxies, 322
Galileo, 177, 327
Gannets, 72
Gases, 158
    heat motion through, 200

Generators, 216
Genes, 90
    movement on chromosomes, 103
Genetics, 28
    beginning of, 88–90
Geological eras, 299–303
Geothermal energy, 229
Germination, 82
Geysers, 299
Gills, 69
Gizzard, 65
Glaciers, 99, 267
Globe, 253
Gold, 157, 169
Gore, Wilbert and Bob, 39
Goretex, 39
Gram, 18
Graphite, 155
Grasshoppers, 66
Gravity
    definition of, 174
    tides and, 313
Great Pyramid, 191
Greenland, 303
Grounding object, 214
Growth, 31
    human, 114
Gulf Stream, 309–310

H

Habitats, 140
Hard copy, 235
Hardware, 236
Healy, Bernadine, 115
Hearing, 109–110
Heart, human, 120–121
    exercise and, 135
Heart disease, 121, 134
Heat, 198–199
    motion of, 199–201
Heat energy, 165, 198–199
    matter states and, 167
Heredity, 88
Highs, 286
High tide, 313
Hiring institutions, 348
Hooke, Robert, 38
Hormones, 114
Host animals, 65

Hot springs, 229
Hubble Space Telescope, 330–331
Human body, 108
    elements and, 39
Humidity, 278
Hummingbirds, 71
Hurricanes, 287
    breaking up, 288
Hybrids, 89
Hydroelectric energy, 228–229

I

Ice, 167, 303
Ice ages, 302
Igneous rocks, 266
Inclined planes, 189
Inertia, 178–179
Infinity, 4
Infrared light, 203
Ingraham, W. James, Jr., 16
Input devices, 235
Insects, 66–67
Insulators
    of electricity, 215
    of heat, 199
Insulin, 92
Invertebrates, 64–67
    in Paleozoic era, 300
Involuntary muscles, 114
Ionosphere, 273
Iron pyrite, 157

J

Jemison, Mae, 326
Joints, 113
Jupiter, 327

K

Keeley, John Worrell, 186
Keyboards, 235
Kilogram, 18
Kilometer, 18
Kinetic energy, 164, 198
Kingdoms, life, 51
    chart, 50
Kurzweil Personal Reader, 6

L

Laboratories, 20
Land breeze, 277

Large intestine, 123
Larynx, 124
Lasers, 5–6, 203
Latitude, 253
Lava, 265
Leaves, plant, 79
Length, metric units of, 18
Levers, 186–187
Life science, 8
    fields of, 28–29
Life span, 33
Light
    definition of, 202–203
    speed of, 203
Light energy, 165
Lightning, 207, 214
Liquids, 158
    heat motion through, 200
Liter, 18
Loads, 184–185
Lodestones. *See* Magnets
Longitude, 253
Low pressure storms, 286–287
Lows, 286
Lubricants, 176
Lungs, human, 124–125

M

McClintock, Barbara, 103
Machines
    compound, 191
    definition of, 184
    first, 191
    simple, 186–190
Magma, 264
Magnetic fields, 219
Magnetic tape, 236
Magnetism, definition of, 218
Magnets, 218–219
Main processing units, 234
Malaria, 55
Mammals, 72–73
Mantle, Earth, 250
Maps, 253–254
Mars, 325–326
    exploration of, 328
Mass, 156–157
    gravity and, 174
    weight and, 175

Massey, Walter, 9
Mathematics, 241
Matter
    chemical change in, 167–168
    definition of, 38
    heat energy and, 167
    physical change in, 167–168
    physical science and, 154
    properties of, 156–157
    states of, 157–158
Mauna Kea, 263
Mayoruna Indians, 69–70
Measurement, 17
Mechanical advantage, 185
    levers and, 187
Mechanical energy, 165
Medicine, from frogs, 69–70
Melting, 167
Membrane, cellular, 40
Memory, computer, 235–236
Mendel, Gregor Johann, 88–90
Menopause, 115
Menstruation, 115
Mercury, planet, 325
Mesosphere, 273
Mesozoic era, 301
Metals, 215–216
Metamorphic rocks, 266
Meteorites, 329
    dinosaur extinction and, 99
Meteorology, 289–290
Meteors, 329
Meter, 18
Metric system, 17–19
    conversion chart, 349
    examples of, 19
Microbiology, 28
Microscope, 37
Mid-ocean ridges, 314–315
Migration, 72
Milky Way, 322
Millimeter, 18
Milton, Katherine, 69–70
Minerals, 266
    as nutrients, 133, 355
Mixtures, 159
Molds, 56
    penicillin and, 21

Molecules, 38, 39
    states of matter and, 158
Mollusks, 65
Moneran kingdom, 55–56
Monitors, 234
Moon, Earth's
    landing on, 328
    tides and, 313
Motion, 178–179
    of molecules, 198–199
Mountains
    in Paleozoic era, 301
    plate movement and, 262
    underwater, 314–315
    wind and, 277
Mount Everest, 249, 263
Movement, 30–31
    animals and, 62
Muscle cells, 63
Muscles, 113–114
Mushrooms, 56
Mutations, 91–92
    evolution and, 102–103

N

NASA, 326
National Science Foundation, 9
National Institute of Health, 115
Natural resources, 146–147
Natural selection, 101
Neap tide, 314
Neptune, 327
Nerve cells, 63
Nervous system, 109, 112
Neurons, 112
Neutrons, 156
Newton, Isaac, 178
Nicotine, 134
Nobel Prize, 45, 103
North Atlantic Current, 310
North Pole, 249
Nose, 110
Novello, Antonia Coello, 125
Nuclear energy, 166, 224–226
Nuclear fission, 166, 224–225
Nuclear fusion, 166, 225
Nuclear waste, 226

Nuclei
　in atoms, 155–156
　in cells, 40
　in monera, 55
Nutrients, 133
Nutrition, 133
　food groups, 354

O

Observations, 5
Occluded fronts, 285
Ocean basins, 315
Oceanographers, 307
Oceans
　currents of, 16, 309–311
　definition of, 308
　floor of, 314–315
　resources from, 316
　waves of, 312–313
Offspring, 88
　natural selection and, 101
Open circuit, 218
Orbits, 322
　of Earth, 251
　elliptical, 323
Organisms
　characteristics of, 29–33
　definition of, 28
　extraterrestrial, 29
　groups of, 52, 140–141
　largest, 33
　structure of, 53
Organs, 108–109
Osteoporosis, 113
Ostriches, 71
Output devices, 235
Ovaries
　human, 114
　plant, 81, 83
Overloaded circuit, 218
Oxygen
　from algae, 316
　cells and, 42
　humans and, 125
　mountain climbing and, 273
　plants and, 80
Oxygen and carbon-dioxide cycle, 145–146
Ozone layer, 273

P

Paleontology, 98
Paleozoic era, 300–301
Pangaea, 260
Parasites, 65
Pen computers, 237
Penicillin, 21
Peregrine falcons, 72
Perpetual motion machine, 186
Petals, plant, 81
Photosynthesis, 79, 145–146
Physical science, 8
　definition of, 154
Physics, 154
Pistils, 81
Pitcher plants, 83
Plague, 129–130
Planets, 323–328
Plant kingdom, 76–83
Plants
　breeding of, 93
　cells of, 43
　definition of, 78
　food and, 30
　meat-eating, 83
　movement and, 31
　reproduction and, 31
Plasma, 121
Plastics, 168
Platelets, 122
Plates, Earth, 261
Plate tectonics, 260–261
Plunkett, Roy, 168
Pluto, 328
Polar climate, 291
Poles
　currents and, 311
　Earth's, 249
　magnetic, 219
Pollen, 81–82
Pollination, 82
Pollution
　air, 273
　nuclear, 226
Pompeii, 265
Populations, 140
Potatoes, 79
Potential energy, 164

Precambrian era, 299–300
Precipitation, 144, 279
Pregnancy, 114
Price, James, 169
Prime meridian, 253
Printers, 235
Prisms, 205
Processes, 5
Processing, computer, 234
Producers, 142
Programs, computer, 236
Properties
    definition of, 154
    of matter, 156–157
Proteins, 133
Protist kingdom, 54–55
Protons, 156
Protozoans, 54–55
Puberty, 114
Pulleys, 188

R
Radiation, 201, 275
Radioactive dating, 298
Rain, 144, 279
    weathering and, 267
Rainbows, 205
Rain forests, 80
Rays, light, 202
Recycling, 140, 147
Red blood cells, 121
Reflection, 204
Refraction, 204
Refrigerants, 168
Reproduction, 31
    human, 114–115
Reptiles, 70–71
    in Mesozoic era, 301
Repulsion
    electrical, 213
    magnetic, 219
Resistance force, 184–185
Resources
    energy, 143, 223–229
    natural, 146–147
    ocean, 316
Respiration, 145–146
    cells and, 42
    human, 124–125

    as recycling, 139–140
    yeast and, 57
Respiratory system, 109, 124–125
Response, 32-33
Revolution, 251
Ring of Fire, 265
Rivers, weathering and, 267
Robots, 193
Rocks, 266
    age of, 297–298
Rolling friction, 176
Roots, plant, 78
Rotation, 251
Roundworms, 65
Rubber, 215

S
Safety rules, 20
Salinity, 308
Salt, 134
    oceans and, 308
Satellites
    artificial, 328
    natural, 325
Saturn, 327
Scales, 68–69
Science
    branches of, 8
    careers in, 8–9, 347
    change in, 4–5
    definition of, 3–4
    support of, 7–8
Scientific method, 14–15
Screws, 190
Sea breeze, 277
Sea-floor spreading, 315
Search for Extraterrestrial Intelligence (SETI), 29
Seasons, 252
Seaweed, 316
Sedimentary rocks, 266
Seed plants, 78
Seeds, 81–83
Seismic sea waves, 313
Senses, human, 109–110
Sex cells
    genes and, 91
    human, 114
Shark teeth, 298
Shooting stars. *See* Meteors

Sight, 109
Silver iodide, 288
Simple machines, 186–190
Skeleton, human, 113
Skin, 110–111
    disease and, 130
Skin cells, 63
Sliding friction, 176
Small intestine, 123
Smell, 110
Smoking
    heart disease and, 134
    lung disease and, 125
Snails, 65
Sodium chloride, 308
Software, 236
Solar collectors, 227
Solar energy, 143, 227–228
Solar system, 248, 322–323
Solids, 157
    heat motion through, 199–200
Solutions, 159
Sound
    definition of, 206
    speeds of, 206
    traveling, 197
Sound waves, 109–110
South Pole, 249
Space
    exploration, 328
    heat motion through, 201
Space rovers, 193
Space shuttle, *Endeavor*, 326
Species, definition of, 52
Spectrum, color, 205
Sperm
    genes and, 91
    human, 114
    plant, 81
Sperm cells, 63
Spiders, 67
Spinal cord, human, 112
Spiny-skinned animals, 65
Sponges, 41, 64
Sports, physics and, 179
Spring tide, 313
Sputnik, 328
Stamens, 81
Starfish, 65
Stars, 3, 13

Static electricity, 213
Stems, plant, 78–79
Stomach, 122
Storms, 285–287
Stratosphere, 273
Stratus clouds, 279
Structure, definition of, 53
Sugar, 80
Sun, 13
    burning out of, 303
    distance from earth, 27
    energy and, 143
    plants and, 80
Sunspots, 323
Support system, 113–114
Surface currents, 309–310
Surgeon General of the U.S., 125
Sweat glands, 111
Symbols, chemical, 155
Systems, human, 109

T

Tapeworms, 65
Taste, 110
Technology, 5–7
Teflon, 168
Temperate climate, 292
Temperature, 198–199
    body, 110–111
Tendons, 113
Testes, 114
Thermosphere, 273
Thunder, 207, 214
Thunderheads, 285–286
Thunderstorms, 285–286
Tidal waves. *See* Seismic sea waves
Tides, 313–314
    as energy source, 229
Time, geological, 299
Time zones, 254–255
Tires, recycling, 147
Tissue, 108
Tongue, 110
Tools, scientific, 20–21
Tornadoes, 287
Touch, 110
Trachea, 124
Traits
    definition of, 88
    dominant, 90

environment and, 93
helpful, 102
recessive, 90
Trenches
ocean, 315
plate movement and, 262
Tropical climate, 291
Troposphere, 272
Trough, wave, 202
Trunks, plant, 78
Turbines, 228
Typhoons, 287

U

Ultraviolet light, 203
danger from, 273
Undersea currents, 310–311
Undertow, 312
Units
definition of, 17
metric, 17–19, 349
Universe, 4, 321–322
Uranus, 327
Uterus, 114

V

ccine, 55
oles, 40, 43
ns, 201

83
73
8

nuclear, 226
respiration and, 125
Water
in atmosphere, 278–279
as energy source, 228–229
humans and, 125
as nutrient, 133
on Mars, 328
plants and, 78–80
Water cycle, 144
Watson, James, 45
Wavelength, 202
Waves
energy and, 201–202
light, 202–203
ocean, 312–313
sound, 109–110, 206
Weather
definition of, 284
forecasting, 289–290
Weathering, 267
Wedges, 190
Wegener, Alfred, 261
Weight, 175
metric units of, 18
Wheel and axle, 188–189
White blood cells, 121–122
disease and, 130
Winds
causes of, 275–276
as energy source, 229
systems of, 277
Woods, Granville T., 192
Word processing, 240
Work, definition of, 184
World Trade Center Towers, 175
Worms, 65
in Precambrian era, 300

Y

Yeast, 57
Yucca Mountain, 226

Z

Zoology, 28